T0220701

Calculus for the Ambitious

From the author of *The Pleasures of Counting* and *Naïve Decision Making* comes a calculus book perfect for self-study. It will open up the ideas of the calculus for any 16- to 18-year-old about to begin studies in mathematics, and will be useful for anyone who would like to see a different account of the calculus from that given in the standard texts.

In a lively and easy-to-read style, Professor Körner uses approximation and estimates in a way that will easily merge into the standard development of analysis. By using Taylor's theorem with error bounds he is able to discuss topics that are rarely covered at this introductory level. This book describes important and interesting ideas in a way that will enthuse a new generation of mathematicians.

T. W. KÖRNER is Professor of Fourier Analysis in the Department of Pure Mathematics and Mathematical Statistics at the University of Cambridge. His previous books include *The Pleasures of Counting* and *Fourier Analysis*.

Calculus for the Ambitious

T. W. KÖRNER
Trinity Hall, Cambridge

CAMBRIDGE
UNIVERSITY PRESS

CAMBRIDGE
UNIVERSITY PRESS

University Printing House, Cambridge CB2 8BS, United Kingdom

One Liberty Plaza, 20th Floor, New York, NY 10006, USA

477 Williamstown Road, Port Melbourne, VIC 3207, Australia

314-321, 3rd Floor, Plot 3, Splendor Forum, Jasola District Centre, New Delhi - 110025, India

79 Anson Road, #06-04/06, Singapore 079906

Cambridge University Press is part of the University of Cambridge.

It furthers the University's mission by disseminating knowledge in the pursuit of education, learning and research at the highest international levels of excellence.

www.cambridge.org
Information on this title: www.cambridge.org/9781107686748

© T. W. Körner 2014

This publication is in copyright. Subject to statutory exception and to the provisions of relevant collective licensing agreements, no reproduction of any part may take place without the written permission of Cambridge University Press.

First published 2014
5th printing 2016

A catalogue record for this publication is available from the British Library

ISBN 978-1-107-06392-1 Hardback
ISBN 978-1-107-68674-8 Paperback

Cambridge University Press has no responsibility for the persistence or accuracy of URLs for external or third-party internet websites referred to in this publication, and does not guarantee that any content on such websites is, or will remain, accurate or appropriate.

A mathematics problem paper (Cambridge Scrapbook, 1859).

Bernard of Chartres used to say that we are like dwarfs on the shoulders of giants, so that we can see more than they, and things at greater distance, not by virtue of any sharpness of sight on our part, or any physical distinction, but because we are carried high and raised up by their giant size.

(*John of Salisbury* Metalogicon)

Poetry is learnt by the continual reading of the poets; painting is acquired by continual painting and designing; the art of proof, by the reading of books filled with demonstrations.

(*Galileo* Dialogue Concerning the Two Chief World Systems)

He understands ye several parts of Mathematicks . . . and which is the surest character of a true Mathematicall Genius, learned these of his owne inclination and by his owne industry without a teacher.

(*Newton* Testimonial for Edward Paget)

What one fool can do, another can.

(Ancient Simian Proverb [7])

Contents

Introduction

Over a century ago, Silvanus P. Thompson wrote a marvellous little book [7] entitled

CALCULUS MADE EASY
Being a Very-Simplest Introduction to
Those Beautiful Methods of Reckoning
which Are Generally Called by the
Terrifying Names of the
Differential Calculus and the Integral Calculus

with the following prologue.

> Considering how many fools can calculate, it is surprising that it should be thought either a difficult or a tedious task for any other fool to learn how to master the same tricks. Some calculus-tricks are quite easy. Some are enormously difficult. The fools who write the textbooks of advanced mathematics – and they are mostly clever fools – seldom take the trouble to show you how easy the easy calculations are. On the contrary, they seem to desire to impress you with their tremendous cleverness by going about it in the most difficult way. Being myself a remarkably stupid fellow, I have had to unteach myself the difficulties, and now beg to present to my fellow fools the parts that are not hard. Master these thoroughly, and the rest will follow. What one fool can do, another can.

For a variety of reasons, the first university course in rigorous calculus is often the first course in which students meet sequences of long and subtle proofs. Sometimes the lecturer compromises and provides rigorous proofs only of the easier theorems. In my opinion, there is much to be said in favour of proving every result and much to be said in favour of proving only the hardest results, but nothing whatsoever for proving the easy results and hand-waving for the harder. The lecturer who does this resembles someone who equips themselves for tiger hunting, but only shoots rabbits.

It is hard to learn the discipline of mathematical proof and it is hard to learn the ideas of the calculus. It seems to me, as it does to many other people, that it is possible, at least in part, to separate the two processes. Like *Calculus Made Easy*, this book is about the ideas of calculus and, although it contains a fair number of demonstrations, it contains no formal proofs.

However, Thompson wrote his book for those who use calculus as a machine for solving problems, and this book is written for those who wish, in addition, to understand how the machine works. I hope it will be found useful by able and enquiring school-children who want to see what lies ahead and by beginning mathematics students as supplementary reading. If others enjoy it or find it useful, so much the better – we can neither choose the friends of our children nor the friends of our books. Potential readers are warned that the book gets harder as it goes along and that it requires fluency in algebra.[1] It will not help you to pass exams or to discourse learnedly at the dinner table on the philosophy of the calculus. This is a book written by a professional[2] for future professionals and that is why I have called it *Calculus for the Ambitious*.

When writing this book I had in mind three sorts of users.

(1) *The desert island student.* If you are reading this book without any outside help, please remember Einstein's advice to a junior high school correspondent: 'Do not worry about your difficulties in mathematics. I can assure you mine are still greater.' If you understand everything, I shall be profoundly impressed. If you understand a great deal, I shall be delighted. If you understand something, I shall be content.

(2) *The student following another course.* I hope you find something of interest. Please remember that there are many ways of presenting the material. When my presentation clashes with your main course, either ignore the material or mentally rewrite it in accordance with that course.

(3) *The student with a helpful friend.* I hope that the spirit of Euler hovers over this book. In his autobiographical notes, he records that Johann Bernoulli refused to give him private lessons

> ... because of his busy schedule. However, he gave me far more beneficial advice, which was to take a look at some of the more difficult mathematical books and work through them with great diligence. Should I encounter some objections or difficulties, he offered me free access to him every Saturday afternoon, and he was gracious enough to comment on the collected difficulties, which was done with such ... advantage that, when he resolved

---

[1] Some of the illustrative material requires a knowledge of elementary trigonometry at the level of Exercise 1.5.3.

[2] My son claims that, when he was very young, he asked me what calculus did and I replied that it put bread, butter and jam on our table.

one of my objections, ten others at once disappeared, which certainly is the best method of making good progress in the mathematical sciences.

Few advisers are Bernoullis and even fewer students are Eulers, but, if you can find someone to give you occasional help in this way, there is no better way to learn mathematics.

The contents of this book do not correspond to the recommendations of any committee, have not been approved by any examination board and do not follow the syllabus of any school, university or government education department known to me. When leaving a party, Brahms is reported to have said 'If there is anyone here whom I have not offended tonight, I beg their pardon.' If any logician, historian of mathematics, numerical analyst, physicist, teacher of pedagogy or any other sort of expert picks up this book to see how I have treated their subject, I can only repeat Brahms' apology. This is an introduction and their colleagues will have plenty of opportunities to put things right later.

Since this is neither a textbook nor a reference book, I have equipped it with only a minimal bibliography and index.

Readers should join me in thanking Alison Ming for turning ill-drawn diagrams into clear figures, Tadashi Tokieda, Gareth McCaughan and three anonymous reviewers for useful comments and several Cambridge undergraduates for detecting numerous errors. In addition, I thank members of Cambridge University Press, both those known to me, like Roger Astley and David Tranah (who would have preferred the title *The Joy of dx*), and those unknown, who make publishing with the Press such a pleasant experience.

This book is dedicated to my school mathematics teachers, in particular to Mr Bone, Dr Dickinson and Mr Wynne Wilson, who went far out of their way to help a very erratic pupil. Teachers live on in the memory of their students.

1

Preliminary ideas

1.1 Why is calculus hard?

Mathematicians find mathematics hard and are not surprised or dismayed if it takes them a long time and a lot of hard work to understand a piece of mathematics. On the other hand, most of them would agree that the only reason we find mathematics hard is that we are stupid.

The basic ideas of the calculus, like the basic ideas of the rest of mathematics, are easy (how else would a bunch of apes fresh out of the trees be able to find them?), but calculus requires a lot of work to master (after all, we are just a bunch of apes fresh out of the trees). Here is a list of some of the difficulties facing the reader.

Mathematics is a 'ladder subject'. If you are taught history at school and you pay no attention during the year spent studying Elizabethan England, you will get bad grades for that year, but you will not be at a disadvantage next year when studying the American Civil War. In mathematics, each topic depends on the previous topic and you cannot miss out too much.

The ladder described in this book has many rungs and it will be a very rare reader who starts without any knowledge of the calculus and manages to struggle though to the end. (On the other hand, some readers will be familiar with the topics in the earlier chapters and will, I hope, be able to enjoy the final chapters.) Please do not be discouraged if you cannot understand everything; experience shows that if you struggle hard with a topic, even if unsuccessfully, it will be much easier to deal with when you study it again.[1]

---

[1] I remember being reduced to tears when studying Cartesian tensors for the first time. I cannot now understand how I could possibly have found them difficult.

Mathematics involves deferred gratification. Humans are happy to do *A* in order to obtain *B*. The are less happy to do *A* in order to to do *B*, to do *B* in order to do *C* and then to do *C* in order to obtain *D*. Results in mathematics frequently require several preliminary steps whose purpose may not be immediately apparent. In Chapter 2 we spend a long time discussing the integral and the fundamental theorem of the calculus. It is only at the end of the chapter that we get our first payback in the form of a solution to an interesting problem.

Mathematics needs practice. I would love to write music like Rossini. My university library contains many books on the theory of music and the art of composing, but I know that, however many books I read, I will not be able to write music. A composer needs to know the properties of musical instruments, and to know the properties of instruments you need to play at least one instrument well. To play an instrument well requires years of practice.

Each stage of mathematics requires fluency in the previous stage and this can only be acquired by hours of practice, working through more or less routine examples. Although this book contains some exercises[2] it contains nowhere near enough. If the reader does not expect to get practice elsewhere, the first volume of *An Analytical Calculus* by A. E. Maxwell [5] provides excellent exercises in a rather less off-putting format than the standard 'door stop' text book. (However, any respectable calculus book will do.)

This is a first look and not a complete story. As I hope to make clear, the calculus presented in this book is not a complete theory, but deals with 'well behaved objects' without giving a test for 'good behaviour'. This does not prevent it from being a very powerful tool for the investigation of the physical world, but is unsatisfactory both from a philosophical and a mathematical point of view. In the final chapter, I discuss the way in which the first university course in analysis resolves these problems. I shall refer to the calculus described in the book as 'the old calculus' and to the calculus as studied in university analysis courses as 'the new calculus' or 'analysis'.[3]

D'Alembert is supposed to have encouraged his students with the cry 'Allez en avant et la foi vous viendra' (push on and faith will come to you). My ideal

---

[2] Sketch solutions to most of the exercises can be found on my home page accessible at http://www.dpmms.cam.ac.uk/~twk/. I have marked a few exercises with a ●. These are less central to the exposition. Some of them are quite long or require some thought.

[3] Whenever the reader sees 'well behaved' she can think of the words 'terms and conditions apply' which appear at the end of advertisements. In the 'old calculus' we know that 'terms and conditions apply', but we do not know exactly what they are. In the 'new analysis' they are spelled out in detail.

reader will be prepared to accept my account *on a provisional basis*, but be prepared to begin again from scratch when she meets rigorous analysis.

It is a very bad idea to disbelieve everything that your teachers tell you and a good idea to accept everything that your teachers tell you. However, it is an even better idea to accept that, though most of what you are taught is correct, it is sometimes over-simplified and may occasionally turn out to be mistaken.

The calculus involves new words and symbols. The ideas of the calculus are not arbitrary, but the names given to the new objects and the symbols used are. At the simplest level, the reader will need to recognise the Greek letter δ (pronounced 'delta') and learn a new meaning for the word 'function'. At a higher level, she will need to accept that the names and notations used reflect choices made by many different mathematicians, with many different views of their subject, speaking many languages at many different times over the past 350 years. If we could start with a clean sheet, we would probably make different choices (just as, if we could redesign the standard keyboard, we would probably change the position of the letters). However, we wish to talk with other mathematicians, so we must adopt their language.

My British accent. The theory and practice of calculus are international. The teaching of calculus varies widely from country to country, often reflecting the views of some long dead charismatic educationalist or successful textbook writer. In some countries, calculus is routinely taught at school level whilst, in others, it is strictly reserved for university. Several countries use calculus as an academic filter, a coarse filter in those countries with a strong egalitarian tradition, a fine one in those with an elitist bent.[4]

I was brought up under a system, very common in twentieth-century Western Europe, where calculus was taught as a computational tool in the last two years of school and rigorous calculus was taught in the first year of university. Some of the discussion in the introduction and the final chapter reflects my background, but I do not think this should trouble the reader.

A shortage of letters. The calculus covers so many topics that we run into a shortage of letters. Mathematicians have dealt with this partly by introducing new alphabets and fonts giving us A, a, α, **A**, **a**, $\aleph$, $\mathcal{A}$, A, a, $\mathbb{A}$, $\mathfrak{A}$, $\mathfrak{a}$, ... Rather than learn a slew of new symbols, I think that my readers will prefer to accept

---

[4] In some cultures, passing an examination in calculus is believed to have the same magical effect that passing an examination in Latin is believed to have in England.

that r will sometimes be an integer[5] and sometimes the radius of a circle, and that the same letter will be used to represent different things in different places.

But it is beautiful. Hill walking is hard work, but the views are splendid and the exercise is invigorating. The calculus is one of the great achievements of mankind and one of the most rewarding to study.

1.2 A simple trick

If you ask a mathematician the value of $20\,019 \times 300\,016$, she will instantly reply that 'to the zeroth order' the answer is $6\,000\,000\,000$. By this, she means that $20\,019$ is very close to $20\,000$ and $300\,016$ is very close to $300\,000$ so $20\,019 \times 300\,016$ is very close to $20\,000 \times 300\,000 = 6\,000\,000\,000$.

Now suppose we ask her to be a little more accurate. She will be only a little slower to reply that 'to the first order' the answer is $6\,006\,020\,000$. How did she obtain her answer? In effect, she observed that

$$20\,019 \times 300\,016 = (20\,000 + 19) \times (300\,000 + 16)$$
$$= 20\,000 \times 300\,000 + 19 \times 300\,000 + 16 \times 20\,000 + 19$$
$$\times 16.$$

The first term consists of two very large numbers multiplied together, the second and third terms consist of a small number multiplying a large number and the fourth term consists of two small numbers multiplied together. The first term is thus much bigger than any of the other terms and so, looking only at this term, we obtain

$$20\,019 \times 30\,0016 = (20\,000 + 19) \times (300\,000 + 16)$$
$$\approx 20\,000 \times 300\,000 = 6\,000\,000\,000$$

'to the zeroth order'.[6] The second and third terms are small compared with the first term, but large compared with the fourth term, so, to obtain a more accurate result, we now look at the first three terms to obtain

$$20\,019 \times 300\,016 = (20\,000 + 19) \times (300\,000 + 16)$$
$$\approx 20\,000 \times 300\,000 + 19 \times 300\,000 + 16 \times 20\,000 = 6\,006\,020\,000$$

'to the first order'.

---

[5] An integer is a positive or negative 'whole number'.
[6] The symbol $\approx$ is read as 'is approximately equal to' and in practice means 'this should be pretty close to, but we are not on oath.'

If we do the full calculation, we obtain

$$20\,019 \times 300\,016 = 6\,006\,020\,304,$$

so the 'zeroth order approximation' was pretty good and the 'first order approximation' was excellent.

It is very hard to find new ideas and so, once a mathematician has found a new idea, she tries to push it as far as it will go. Let us look at the structure of our argument. We are interested in multiplying $(a + \delta a)$ by $(b + \delta b)$, where δa and δb are very small compared to a and b.

Notational remark. By long tradition, δa (pronounced 'delta a') is a *single symbol*[7] and should be thought of as 'a little bit of a'. American texts often use Δa rather than δa for the same purpose.[8]

We observe that

$$(a + \delta a) \times (b + \delta b) = a \times b + \delta a \times b + \delta b \times a + \delta a \times \delta b.$$

The first term is much bigger than any of the other terms and so, looking only at this term, we obtain

$$(a + \delta a) \times (b + \delta b) \approx a \times b$$

'to the zeroth approximation'. The second and third terms are small compared with the first term, but large compared with the fourth term, so to obtain a more accurate result, we now look at the first three terms to obtain

$$(a + \delta a) \times (b + \delta b) \approx a \times b + \delta a \times b + \delta b \times a.$$

'to the first approximation'.

Looking at the algebra, we see that our argument does not depend on the sign of the various terms, but simply on their magnitude.

Exercise 1.2.1. *Use this insight to find*

$$20\,019 \times 299\,987$$

'to the zeroth and first order'. Find the result exactly for the purposes of comparison.

We can push things a little further. One obvious direction is to try multiplying three terms $a + \delta a$, $b + \delta b$ and $c + \delta c$, where δa, δb and δc are very small

---

[7] In particular, it does not mean $\delta \times a$.
[8] In the Greek alphabet, Δ is the capital version of δ so, conveniently, Δa and δa have the same pronunciation.

compared to a, b and c. We observe that

$$(a + \delta a) \times (b + \delta b) \times (c + \delta c)$$
$$= a \times b \times c + \delta a \times b \times c + \delta b \times a \times c + \delta c$$
$$\times b \times a + \delta a \times \delta b \times c + \delta a \times \delta c \times b + \delta b \times \delta c \times a + \delta a \times \delta b \times \delta c.$$

The first term is much bigger than any of the other terms and so, looking only at this term, we obtain

$$(a + \delta a) \times (b + \delta b) \times (c + \delta c) \approx a \times b \times c$$

'to the zeroth order'. The next three terms are small compared with the first term, but large compared with the remaining terms, so, to obtain a more accurate result, we now look at the first three terms to obtain

$$(a + \delta a) \times (b + \delta b) \times (c + \delta c) \approx a \times b \times c + \delta a \times b \times c + \delta b \times a \times c$$
$$+ \delta c \times b \times a$$

'to the first order'.

It is now natural to observe that the next three terms are small compared to the previous terms, but large compared to the final term, so

$$(a + \delta a) \times (b + \delta b) \times (c + \delta c)$$
$$\approx a \times b \times c + \delta a \times b \times c + \delta b \times a \times c + \delta c$$
$$\times b \times a + \delta a \times \delta b \times c + \delta a \times \delta c \times b + \delta b \times \delta c \times a$$

'to the second order'. However, for the moment, we shall confine ourselves to zeroth order and first order approximations.[9]

Exercise 1.2.2. *Suppose that δx is small in magnitude compared with x. Find $(x + \delta x)^3$ to the first order.*

Suppose that n is a positive integer. Find $(x + \delta x)^n$ to the first order, giving reasons for your answer.

What happens if we consider other operations? Addition is very simple, since

$$(a + \delta a) + (b + \delta b) = a + b + (\delta a + \delta b),$$

so we have the unremarkable observation that

$$(a + \delta a) + (b + \delta b) \approx a + b$$

---

[9] At some point, the reader should observe that two zeroth order estimates of the same quantity may differ to the first order, two first order estimates of the same quantity may differ to the second order and so on.

to the zeroth order and

$$(a + \delta a) + (b + \delta b) \approx a + b + (\delta a + \delta b),$$

to the first order.

Exercise 1.2.3. *Make a similar observation for* $(a + \delta a) - (b + \delta b)$.

What about

$$\frac{b + \delta b}{a + \delta a}\ ?$$

Note first that we know how to deal with multiplication, so it suffices to look at the simpler problem of evaluating

$$\frac{1}{a + \delta a}$$

approximately. If we stare at the problem long enough, the following idea may occur to us. Let us write

$$u = \frac{1}{a} \text{ and } u + \delta u = \frac{1}{a + \delta a}.$$

If δa is small in magnitude compared to a, then $a + \delta a$ will be close to a and so $u + \delta u$ will be close to u. In other words, δu will be small in magnitude compared to u. Thus, working to first order,

$$1 = (a + \delta a) \times (u + \delta u) \approx a \times u + a \times \delta u + u \times \delta a$$
$$= 1 + a \times \delta u + u \times \delta a.$$

Subtracting 1 from both sides and rearranging, we get

$$a \times \delta u \approx -u \times \delta a,$$

so

$$a \times \delta u \approx -\frac{1}{a} \times \delta a,$$

that is to say,

$$\delta u \approx -\frac{1}{a^2} \times \delta a$$

to the first order.

Combining the results of the last paragraph with our previous results on multiplication, we see that, if δa and δb are small in magnitude compared with *a* and *b*,

$$\frac{b + \delta b}{a + \delta a} = (b + \delta b) \times (u + \delta u) \approx b \times u + b \times \delta u + u \times \delta b$$

$$\approx \frac{b}{a} - \frac{b}{a^2} \times \delta a + \frac{1}{a} \times \delta b$$

to the first order.

The same kind of idea can be used to estimate $\sqrt{a + \delta a}$ when *a* is positive and large in magnitude compared with δa. Let us write

$$v = \sqrt{a} \text{ and } v + \delta v = \sqrt{a + \delta a}.$$

If δa is small in magnitude compared to *a*, then $a + \delta a$ will be close to *a* and so $v + \delta v$ will be close to *v*. In other words, δv will be small in magnitude compared to *v*. Thus, working to first order,

$$a + \delta a = (v + \delta v)^2 \approx v^2 + 2v \times \delta v = a + 2\sqrt{a} \times \delta v.$$

Subtracting *a* from both sides and rearranging, we get

$$\delta a \approx 2\sqrt{a} \times \delta v,$$

so

$$\delta v \approx \frac{1}{2\sqrt{a}} \times \delta a$$

to the first order.

Exercise 1.2.4. *Estimate* $\sqrt{1\,000\,003}$ *and compare your estimate with the answer given by your calculator.*

Exercise 1.2.5. *We write* $a^{1/3}$ *for the cube root of a. Find a first order estimate for* $(a + \delta a)^{1/3}$ *when a is large in magnitude compared with* δa. *Estimate* $1\,000\,003^{1/3}$ *and compare your estimate with the answer given by your calculator.*

We can obtain many other results by combining the ones we already have.

Example 1.2.6. *Heron's formula states that the area A of a triangle whose sides have lengths a, b and c is given by*

$$A = \sqrt{s(s - a)(s - b)(s - c)}, \text{ where } s = \frac{a + b + c}{2}.$$

Use the formula to estimate the difference in area between a triangle whose sides have lengths a, b and c and a triangle whose sides have lengths $a + \delta a$, $b + \delta b$ and $c + \delta c$, where, as usual, δa, δb and δc are small in magnitude compared with a, b and c.

Solution. Let us write $u = s - a$, $v = s - b$, $w = s - c$ and $T = suvw$ and let us denote the corresponding quantities for the new triangle by $u + \delta u$, $v + \delta v$, $w + \delta w$ and $T + \delta T$. The quantities $A + \delta A$ and $s + \delta s$ will have the obvious meanings.

Our earlier result on square roots tells us that

$$A + \delta A = \sqrt{T + \delta T} \approx \sqrt{T} + \frac{1}{2\sqrt{T}}\delta T = A + \frac{1}{2A}\delta T$$

so that

$$\delta A \approx \frac{1}{2A}\delta T$$

to the first order. A simple modification of our rule concerning products of two and three objects gives[10]

$$T + \delta T = (s + \delta s)(u + \delta u)(v + \delta v)(w + \delta w)$$
$$\approx suvw + uvw\delta s + svw\delta u + suw\delta v + suv\delta w$$
$$= T + uvw\delta s + svw\delta u + suw\delta v + suv\delta w,$$

so that

$$\delta T \approx uvw\delta s + svw\delta u + suw\delta v + suv\delta w$$

or, more neatly,

$$\delta T \approx T \left(\frac{\delta s}{s} + \frac{\delta u}{u} + \frac{\delta v}{v} + \frac{\delta w}{w} \right)$$

to the first order.

Combining our two results gives

$$\delta A \approx \frac{1}{2A} \times T \left(\frac{\delta s}{s} + \frac{\delta u}{u} + \frac{\delta v}{v} + \frac{\delta w}{w} \right) = \frac{\sqrt{A}}{2} \left(\frac{\delta s}{s} + \frac{\delta u}{u} + \frac{\delta v}{v} + \frac{\delta w}{w} \right)$$

to the first order. This result is nice and symmetric, but the reader may object that, for example, δs depends on δa, δb and δc. She is invited to do the next exercise.

Exercise 1.2.7. *We use the notation of the last example. Show that*

$$\delta u = \frac{\delta b + \delta c - \delta a}{2}$$

and write down a first order formula for δA in terms of a, b, c, δa, δb and δc.

---

[10] Note that, having established the convention that δa is a single symbol, we have reverted to the standard practice of writing $a \times b = ab$ and $x \times \delta y = x\delta y$.

I suspect that your formula will not be particularly pretty. Write down a more elegant first order formula for δA in terms of u, v, w, δu, δv and δw.

The reader may ask if the ideas of this section are useful or merely some kind of party trick. She should note that many engineers and physicists are wizards at this sort of calculation, which strongly suggests that they find it useful.

Engineers ask 'what happens if I tweak this process?'. Just as the formula we discussed shows what happens to the area of a triangle if we make a small change in the lengths of the sides, so engineers can make 'back of an envelope' calculations about what happens to an output when they make small changes in the inputs.

Physicists know that all measurements have errors. They ask how much those errors affect the final result. The formula just obtained shows how small errors in measurement in the lengths of the sides would change the calculated area and the same technique enables physicists to make 'back of an envelope' calculations about the cumulative effects of different errors in measurement.

It is clear that what I have presented here is more of an art than a science. As the reader learns more about the calculus, she will find that there are relatively simple rules (for example, the 'function of a function' rule which we meet later) that cover many of the circumstances when we wish to perform approximate calculations, but sometimes the only tool available will be her own unaided ingenuity.

A more basic problem is 'how small is small?'. How small do the 'delta quantities' have to be relative to the 'large quantities' for first order calculations to be useful? Mathematicians have developed heavy mathematical machinery (for example, Taylor's theorem, which we meet later) for the purpose of answering this question, but most physicists and engineers rely on *experience* rather than *theory* to decide 'how small the deltas need to be' and do so very successfully.[11]

Advice. In order to gain the experience required, cultivate the habit, whenever you meet an interesting formula, of calculating the effect of small changes in the variables to first order and then thinking how small the changes need to be for your calculation to be useful.

Notational remark. We have used the approximation symbol in approximate equations like

$$\delta x \approx \delta y \text{ to first order.}$$

---

[11] Any fool can build something which will last forever and cost a thousand dollars. An engineer can build something that lasts as long as is required and costs a hundred dollars.

It is more usual to write

$$\delta x = \delta y \text{ to first order}$$

and we shall follow this convention from now on.

Exercise 1.2.8. *If $b^2 > 4c$, the quadratic equation $x^2 + bx + c = 0$ has two roots[12]*

$$\frac{-b + \sqrt{b^2 - 4c}}{2} \text{ and } \frac{-b - \sqrt{b^2 - 4c}}{2}.$$

What is the effect of small changes in b and c?

Exercise 1.2.9. [Ring round the moon] *For the purposes of this question, we suppose the moon to be a perfect sphere of radius 1731 kilometres.*

The International Space Corps decides to build a fence along the equator of the moon. The top of the fence will be a continuous steel bar (bent into circular shape) at a fixed distance from the moon's surface. It is intended that the top of the fence will be 1 metre above the surface, but, owing to a mistake in calculation, the bar is 10 metres too long. How much higher will the top of the fence have to be?

[Moral: we need to think carefully about what is small and what is not.]

Exercise 1.2.10. *Shares in Pumpkin Computers increase in value by 1% (i.e. their value is multiplied by 1.01) and then fall by 1% (i.e. their value is multiplied by 0.99). This happens 40 times. Is it true that their value is essentially unchanged?*

Shares in Melon Computers increase in value by 40% (i.e. their value is multiplied by 1.4) and then fall by 40% (i.e. their value is multiplied by 0.6). Is it true that their value is essentially unchanged? Comment.

1.3 The art of prophecy

Mathematicians do mathematics because they enjoy it. People pay mathematicians to do mathematics because they believe mathematics to be useful. One of the things that mathematics is believed to be useful for is predicting the future. How much will this bridge cost to build? What is the weather going to be like tomorrow? How many car accidents will there be next year involving drivers insured with a particular company and how much will that company have to pay out?

---

[12] This book only deals with real numbers, so 'root' always means real root.

Given the demand for prophecy, how should we go about satisfying it? We should bear in mind Bacon's counsel in his *Advancement of Learning*; 'If a man will begin with certainties, he shall end in doubts; but if he will be content to begin with doubts, he shall end in certainties.' In keeping with this advice, we shall consider only very simple prophecy in very simple cases.

Our discussion involves the idea of a function $f(t)$ (pronounced 'f of t'). The reader should think of $f(t)$ as the temperature at the time t or the height of an aircraft at time t. We shall talk about 'the function f of the *variable t*'.[13] In advanced work, it is more usual to talk about 'the function f' without referring to t and we shall do this more frequently as the book progresses. Sometimes we can write down an explicit expression for $f(t)$, for example, $f(t) = t^2 + 3$ and sometimes (as with temperature) we cannot.

We will make much use of the modulus function $m(t)$ defined by

$$m(t) = \begin{cases} -t & \text{if } t < 0, \\ t & \text{if } t \geq 0. \end{cases}$$

We write $|t| = m(t)$ and call $|t|$ the *absolute value* of t.

Exercise 1.3.1. *Sketch the graph of the modulus function.*

Exercise 1.3.2. *Let a, b and c be given numbers. By considering the possible cases, demonstrate the following results which will be used on almost every page of this book.*

(i) $|a + b| \leq |a| + |b|$.
(ii) $|ab| = |a||b|$.

Use (i) to show that

(iii) $|a - c| \leq |a - b| + |b - c|$.
(iv) *Give an example of distinct a, b, c such that* $|a - c| = |a - b| + |b - c|$.

Give an example of distinct a, b, c such that $|a - c| < |a - b| + |b - c|$.

You should think of $|a - b|$ as 'the distance from a to b'. Part (iii) of the previous exercise thus says that 'the distance from a to c is less than or equal to the distance from a to b plus the distance from b to c' or 'the distance from a to c via b is never less than the direct distance from a to c'.

---

[13] The reader may ask whether the notion of function extends to 'functions $f(x, y)$ depending on two variables x and y'? The answer is yes, as we shall see in Chapter 8, but it is better to master walking before trying to run.

The simplest version of the prophecy problem asks us to predict $f(t + h)$ knowing the value of $f(t)$. Thus, knowing the temperature $f(t)$ at time t (now), we want to guess the temperature $f(t + h)$ at some future time $t + h$.

It is, almost always, an excellent first try at prophecy to predict that 'tomorrow will be like today'. In the case of temperature, this prediction is often false when taken literally, but, if we predict that the temperature in one minute's time will be the same as that now, we are very unlikely to be wrong by very much. In other cases, we can safely make longer term predictions. If I do not cut my hair, the length of my hair tomorrow is likely to be pretty close to its length today (though, of course, the prediction will not hold over two months). The distance between a particular landmark in Scotland and a particular landmark in Newfoundland is likely to be pretty much unchanged in a year's time (though, of course, there may be big changes over a few tens of millions of years).

We shall call functions where 'tomorrow will be much like today' *over all sufficiently short time scales* continuous functions. More exactly, we shall call a function f continuous if $f(t + h)$ is guaranteed to be as close as we choose to specify to $f(t)$ provided that h is sufficiently small. (Here and elsewhere, we allow h to be both positive and negative and the statement 'h small' means '$|h|$ small'. Thus we also require that 'yesterday was much like today'.)

In advanced courses, this notion is made completely precise by using the following form of words.

Definition 1.3.3. *A function f is continuous at t if, given any $u > 0$, we can find a $v > 0$ such that $|f(t + h) - f(t)| \leq u$ whenever $|h| \leq v$.*

However, the beginner will find it more profitable to think of a continuous function as a function which is 'close to constant at all sufficiently fine scales'.

A little philosophy. It is easy to write down functions which are continuous, for example, the constant function given by $f(t) = a$ for every t. It is almost as easy to write down functions which are not continuous (or, as we shall say, *discontinuous*). A simple example is the *Heaviside function* H defined by

$$H(t) = \begin{cases} 0 & \text{if } t < 0, \\ 1 & \text{if } t \geq 0. \end{cases}$$

Exercise 1.3.4. *Sketch the graph of the Heaviside function.*

Since the time of the Ancient Greeks, philosophers have argued over whether everything in the physical world is continuous or nothing is. If the reader is asked to give an example of a discontinuous phenomenon, she might choose the bursting of a dam or the explosion of a shell. (It is clear why we associate discontinuity, that is to say, the failure of things to remain roughly as they are over a reasonable length of time, with catastrophe.) The philosopher who believes that everything is continuous would point out that, if we make a high speed film of the bursting of a dam or the explosion of a shell and then run it at normal speed, we see events unfolding in a natural, continuous manner. The apparent discontinuity is the result of viewing things at an insufficiently fine scale. Everything is continuous.

In reply, the proponents of discontinuity would ask us to look at a 'continuous' piece of rope. We know that the rope is actually composed of separate fibres, that those fibres are composed of separate molecules and that those molecules are in turn composed of atoms. A simple picture of those atoms shows an object consisting mainly of empty space with distant electrons orbiting a tiny nucleus and, if we seek to look further, the familiar notions of space and time begin to break down. Everything is discontinuous.

In practice, physicists and engineers avoid taking part in this controversy by observing that every investigation involves working to a certain scale. To a geologist, a hundred million years is a long time but a century is an instant. To the botanist, a year is a long time but a minute is an instant. To the explosives expert, five seconds is a long time but a millionth of a second is an instant. The correct question is not 'is this function continuous in the mathematical sense?' but 'can we usefully assume that the function is continuous?' that is to say, 'do small changes produce small effects *at the scale we are considering*?'. The skill of the pioneering engineer or physicist consists in choosing the right scale and then correctly identifying those things which are predictable at that scale.[14] As the great physicist Maxwell wrote, when still a student,

> ...there is a tendency either to run together into masses or to split up into limbs. The dimmed outlines of phenomenal things all merge into another unless we put on the focusing glass of theory and screw it up sometimes to one pitch of definition, and sometimes to another, so as to see down into different depths through the great millstone of the world.
>
> Campbell and Garnett *Life of Maxwell*

We know that currency is discontinuous. To millionaires, this does not matter and they can view their fortune as a continuous function of time. The very poor cannot take the same view, since the gain or loss of one unit of currency is an important event for them.

The expectation that, over a sufficiently short time, things remain almost unchanged is probably hardwired into our brains. Psychologists have observed

---

[14] In the case of numerical weather forecasting, it was discovered that the appropriate time scale was a hundred times finer than we might expect.

that even very young babies show surprise when objects are made to disappear. So far as this book is concerned, this has the advantage that I can safely claim that certain things (like the fact that the sum of two continuous functions is continuous) are obvious. However, our intuition resembles the physicist's intuition in that it deals with things which remain almost unchanged over short periods *at a particular scale*. The mathematical definition demands that a continuous function remains almost unchanged *at every sufficiently fine scale* and this stronger demand turns out to have some rather subtle consequences. If the reader goes on to study mathematics at a higher level, she will find that it requires a fair amount of work to bypass our fallible intuition.[15]

1.4 Better prophecy

Although the prediction that tomorrow will be like today is a very good one (once we have found the appropriate scale), many non-mathematicians believe that they can make such a prediction for themselves without needing the services of a mathematician. It is also noticeable that things *do* change. Can we improve our methods to take account of these two facts?

One possibility is suggested by the artillery technique of 'bracketing', in which the first round fired overshoots the target and the second undershoots. After observing the amount of the over and undershoot, the gunner tries to adjust the angle of elevation so that the third shot hits the target. Here is a fictitious example which the reader should think about before proceeding.

Exercise 1.4.1. *For the first shot, the angle of elevation is* 32 *degrees and the shell lands* 50 *metres behind the target. For the second shot, the angle of elevation is* 31 *degrees and the shell lands* 100 *metres in front of the target. What angle of elevation should be chosen for the third shot?*

If, as I hope she does, the reader chooses the angle $31\frac{2}{3}$ degrees, she will, consciously or unconsciously, be following the principle that 'small changes look linear'. If $f(t)$ is the distance (in metres, say) that the shell travels from the gun when the angle of elevation is t, then there is a constant A such that

$$f(t + h) \approx f(t) + Ah$$

provided that h is small.

---

[15] In effect, the first university lecture she attends on the topic will be devoted to explaining why certain things which she thought were obvious are not. The next three lectures will be devoted to showing that these non-obvious things are indeed the case. Many people find this way of proceeding ridiculous. They are not mathematicians.

Exercise 1.4.2. *Suppose that it is* exactly *true that the range of the gun in Exercise 1.4.1 is given by* $f(31 + h) = f(31) + Ah$. *Find A and show that, if we choose* $h = \frac{2}{3}$, *we will hit the target.*

Important. The next paragraph is very important and the reader should reflect on it now and in the future.

What do we mean when we say $f(t + h) \approx f(t) + Ah$? We do not mean simply that $f(t + h)$ is close to $f(t) + Ah$, since all that says is that f is continuous.[16] In order that our 'bracketing technique' should work, we must mean that $f(t) + Ah$ is close to $f(t + h)$ *in comparison with* $|h|$, that is to say, we mean that

$$\frac{f(t + h) - \big(f(t) + Ah\big)}{h}$$

is small. More exactly, we mean that we can make

$$\frac{f(t + h) - \big(f(t) + Ah\big)}{h}$$

as small as we like, provided that we take h sufficiently small.

If this is the case, we say that f is *differentiable* and write $A = f'(t)$ (pronounced 'f dashed of t' or 'f prime of t'). We call $f'(t)$ the *derivative* of f at t. Sometimes we shall write

$$f(t + h) = f(t) + f'(t)h + o(h).$$

Mathematicians pronounce '$+o(h)$' as 'plus little o of h' but I *very strongly recommend* that the reader pronounces it as

'plus an error term which diminishes faster than linear'

or

'plus an error term which diminishes faster than $|h|$'.

In advanced courses, the notion of differentiability is made completely precise by using the following form of words.

Definition 1.4.3. *A function f is differentiable at t with derivative $f'(t)$ if, given any $u > 0$, we can find a $v > 0$ such that*

$$|f(t + h) - f(t) - f'(t)h| \le u|h|$$

whenever $|h| \le v$.

---

[16] If h is small, then $f(t + h)$ is close to $f(t) + Ah$ when $f(t + h)$ is close to $f(t)$ and vice versa.

However, the beginner will find it more profitable to think of a differentiable function as a function which 'looks linear at any sufficiently fine scale'.

The reader will observe that, if f is differentiable, then, in the language of Section 1.2,

$$f(t + \delta t) = f(t) + f'(t)\delta t$$

to first order. In Section 1.2, we deliberately left the meaning of the words 'to first order' rather vague, but Definition 1.4.3 is a precise test for differentiability.

The operation of finding f' from f is called 'differentiating f' or 'finding the derivative of f'.

If $f(t)$ is the height of an object in metres at a time t seconds, $f'(t)$ will be measured in metres per second. It seems reasonable to think of $f'(t)$ as a 'rate of change'.

Exercise 1.4.4. *Convince yourself of the truth of the previous paragraph by thinking about the special case when $f(t) = At + B$ (so that f is exactly linear).*

The alert reader may feel that, if linear approximation is good, then quadratic approximation will be better and cubic approximation even better. There are two reasons, both important, why we do not follow this path. The first is that simple tools are less likely to fail in our hands than complicated ones. It is a good rule to use simple methods which you understand, rather than apparently better methods which you do not. The second is one of the most remarkable things about the calculus. It turns out that results about these higher order approximations can be deduced from the first order case! We shall see why this is so in Section 6.2.

The derivatives of many common functions can be obtained by using some simple rules. In order to understand these rules we first look at the simple case where f is *exactly* linear, that is to say,

$$f(x) = ax + p,$$

where a and p are constants.

Suppose that $f(x) = ax + p, g(x) = bx + q$. Then $f(t + h) = f(t) + ah$, $g(t + h) = g(t) + bh$ and we have the following rules.

(1)′ If we write $u_1(x) = f(x) + g(x)$, then

$$u_1(t + h) - u_1(t) = (a + b)h$$

so $u_1'(t) = f'(t) + g'(t)$.

(2)′ If we write $u_2(x) = f(x) \times g(x)$, then

$$u_2(t + h) - u_2(t) = (f(t) + ah) \times (g(t) + bh) - f(t) \times g(t)$$
$$= (ag(t) + bf(t))h + abh^2 = (ag(t) + bf(t))h + o(h)$$

so $u_2'(t) = ag'(t) + bf'(t) = f(t)g'(t) + g'(t)f(t)$.

(3)′ If we write $u_3(x) = f(g(x))$, then

$$u_3(t + h) - u_3(t) = f(g(t + h)) - f(g(t))$$
$$= f(g(t) + bh) - f(g(t)) = abh$$

so $u_3'(t) = ab = f'(g(t))g'(t)$ (the reason why we have chosen this form will appear in a moment).

We now look at the more general case. We assume that things are *well behaved* and what looks as though it ought to be the case is the case. Suppose that f and g are differentiable. Then

$$f(t + h) = f(t) + f'(t)h + o(h), \ g(t + h) = g(t) + g'(t)h + o(h)$$

and we have the following rules.

(1)″ If we write $u_1(x) = f(x) + g(x)$, then

$$u_1(t + h) - u_1(t) = (f'(t) + g'(t))h + o(h)$$

so $u_1'(t) = f'(t) + g'(t)$.

(2)″ If we write $u_2(x) = f(x) \times g(x)$, then

$$u_2(t + h) - u_2(t) = (f(t) + f'(t)h + o(h)) \times (g(t) + g'(t)h + o(h))$$
$$- f(t) \times g(t)$$
$$= (f'(t)g(t) + g'(t)f(t))h + o(h)$$

so $u_2'(t) = f(t)g'(t) + g'(t)f(t)$.

If the reader tries to use *exactly* the same argument in the third case, it will fail. A little reflection shows that we are still interested in the rate of change of g at t, but now *it is the rate of change of f at $g(t)$ which is important*. We observe that $f(g(t) + k) = f(g(t)) + f'(g(t))k + o(k)$, $g(t + h) = g(t) + g'(t)h + o(h)$ and we have the following rule.

(3)″ If we write $u_3(x) = f(g(x))$, then

$$u_3(t + h) - u_3(t) = f(g(t + h)) - f(g(t))$$
$$= f(g(t) + g'(t)h + o(h)) - f(g(t)) = f'(g(t))g'(t)h + o(h),$$

so $u_3'(t) = ab = f'(g(t))g'(t)$.

These rules are usually written in a more compressed form.

(1) **[Addition rule]** $(f + g)'(t) = f'(t) + g'(t)$.
(2) **[Product rule]** $(f \times g)'(t) = f'(t)g(t) + f(t)g'(t)$.

To avoid confusion with multiplication, we introduce the notation $f \circ g(x) = f(g(x))$. Our third rule is then written as follows.

(3) **[Function of a function rule]**[17] $(f \circ g)'(t) = g'(t)(f' \circ g)(t)$.

You are, of course, free to join most mathematicians and just write $(fg)'(t) = g'(t)f'(g(t))$.

Exercise 1.4.5. *(i) If $g(t) = a$ (that is to say, g is the constant function with value a), show that $g'(t) = 0$.*
  (ii) Use the product rule to show that, if $u(t) = af(t)$, then $u'(t) = af'(t)$.
  (iii) If $g_1(t) = t$, verify that $g_1'(t) = 1$.
  (iv) Use the product rule to show that, if $g_2(t) = t^2$ (that is to say, $g_2(t) = g_1(t) \times g_1(t)$), then $g_2'(t) = 2t$.
  (v) If $g_3(t) = t^3$ (so $g_3(t) = g_1(t) \times g_2(t)$), use the product rule to show that $g_3'(t) = 3t^2$.
  (vi) By repeating the type of argument just suggested, show that, writing $g_m(t) = t^m$, we have $g_n'(t) = ng_{n-1}(t)$ for all integers $n \geq 1$.
  (vii) By observing that $g_n(t)g_{-n}(t) = 1$ and applying the product rule, show that

$$g'_{-n}(t) = -\frac{n}{t^{n+1}}$$

for all integers $n \geq 1$. Conclude that $g_n'(t) = ng_{n-1}(t)$ for all integers $n \neq 0$. What happens if $n = 0$?
  (viii) If

$$P(t) = a_n t^n + a_{n-1}t^{n-1} + \cdots + a_0,$$

use our differentiation rules and the results of this question to find $P'(t)$.
  (ix) **[Quotient rule]** *By applying the function of a function rule to $g_{-1} \circ f$, show that, if f is well behaved and $h(t) = 1/f(t)$, then $h'(t) = -f'(t)/f(t)^2$.*
  (x) Obtain the result of (ix) directly by applying the product rule to the equation $1 = f(t)h(t)$.
  (xi) If f and g are well behaved and $h(t) = f(t)/g(t)$, express h' in terms of f, g and their derivatives.

There is a further rule which requires a new definition.

---

[17] Also called the *chain rule*.

Definition 1.4.6. *If $f \circ g(x) = x$ for all values of x and $g \circ f(t) = t$ for all values of t, we say that g is the* inverse function *of f and write $f^{-1}(x) = g(x)$.*

The inverse function *only exists in special cases*, but, when it does exist, it is very useful. We pronounce f^{-1} as 'f inverse'.

Observe that, *if* the inverse function f^{-1} of f exists and both f and f^{-1} are well behaved, then, writing $h(x) = x$, we have

$$h(x) = f \circ f^{-1}(x)$$

and so, using the function of a function rule, we have

$$1 = h'(x) = (f \circ f^{-1})'(x) = (f^{-1})'(x)f'\big(f^{-1}(x)\big)$$

and division gives our fourth rule.

(4) **[Inverse function rule]** If f^{-1} exists and is well behaved,[18] then

$$(f^{-1})'(x) = \frac{1}{f'\big(f^{-1}(x)\big)}.$$

The next exercise shows how the inverse function rule can be used.

Exercise 1.4.7. *(i) Recall that, if p is an integer with $p \geq 1$ and $x > 0$, we write $x^{1/p}$ for the pth root of x. If we write $g_{1/p}(t) = t^{1/p}$, use the inverse function rule to show that*

$$g'_{1/p}(t) = \frac{1}{pt} g_{1/p}(t).$$

(ii) Let us write $g_{q/p}(t) = (g_{1/p}(t))^q = t^{q/p}$, where p and q are integers with $p \geq 1$ and $q \neq 0$. Use the function of a function rule to show that

$$g'_{q/p}(t) = \frac{q}{p} g_{(q/p)-1}(t).$$

The reader will observe that our calculations with derivatives echo our calculations 'to first order' in Section 1.2. The use of the rules above, together with the observation that

$$f(x + \delta x) = f(x) + f'(x)\delta x$$

to first order, will often remove or, at least, reduce the need for ingenuity in computing quantities to first order.

---

[18] When she studies rigorous calculus, the reader will be given a proof along rather different lines which shows (what we shall only assume) that, if f is well behaved, then f^{-1} is. There will also be a discussion of the conditions required for an inverse to exist and the 'range' over which it will be defined.

In this book we will need to differentiate several complicated-looking functions. In each case we can solve the complicated-looking problem by splitting it into simpler sub-problems and using the rules found in this section. The next exercise gives an example of the procedure.

Exercise 1.4.8. *(i) Let*

$$f(x) = \frac{1 + x^{1/3}}{1 + \sqrt{1 + x^2}}.$$

Our object is to find $f'(x)$.
  Observe that $f(x) = g(x)/h(x)$ with

$$g(x) = 1 + x^{1/3}, \ h(x) = 1 + \sqrt{1 + x^2}.$$

Now note that $g(x) = a(x) + b(x), h(x) = a(x) + c(x)$ with

$$a(x) = 1, \ b(x) = x^{1/3}, c(x) = \sqrt{1 + x^2}$$

and $c(x) = S\big(u(x)\big)$ with

$$u(x) = 1 + x^2, \ S(x) = \sqrt{x}.$$

Now obtain $a'(x), S'(x), u'(x), c'(x), h'(x), b'(x), g'(x)$ and $f'(x)$.
  (ii) Write down some more functions along the lines laid out in (i) and find their derivatives.

With practice and experience, the reader will find that she can reduce the number of sub-problems required.

On remembering and understanding. Chess players do not carry around a notebook explaining how the knight moves and bridge players do not clutch a memo reminding them that there are four suits each of 13 cards. In the same way, mathematicians do not consult a 'formula book' for the differentiation rules just given (or for anything else). It is impossible to work on difficult mathematics unless you can work quickly and efficiently through the easy parts.
  When a mathematician cannot remember a fact or a formula, her first action is to attempt to re-derive it for herself. If she cannot do this, she concludes that she does not understand the result and looks up *not the result* but its *derivation* and studies the derivation until she is certain that she understands why the result is true. If you understand why a result is true, it is easy to remember it. If you do not understand why a result is true, it is useless to memorise it.[19]

---

[19] Except the night before an examination.

There is a real danger that one becomes so accustomed to using a result like the function of a function rule that one forgets why it is true. (This is one of the causes of 'blackboard blindness' where a lecturer suddenly finds herself unable to prove a simple result.) I advise you to stop and consider why a result is true every tenth time you use it.

A little philosophy. We have now seen several mathematical functions that are differentiable. Here is an example of a function which is not differentiable at a particular point.

Exercise 1.4.9. *Let* $m(t) = |t|$. *Show that, if* $m(h) - m(0) = ah + o(h)$ *for* $h > 0$, *then* $a = 1$. *Show that, if* $m(h) - m(0) = bh + o(h)$ *for* $h < 0$, *then* $b = -1$. *Deduce that there does not exist a c with* $m(h) - m(0) = ch + o(h)$ *when h is allowed to be both positive and negative. Thus m is not differentiable at the point* $t = 0$.

If we look at the real world, the correct question to ask is not whether 'real world functions' are differentiable, but 'is a certain process usefully represented by a differentiable function?' In the nineteenth century it was believed that the answer must always be yes, but the twentieth century provided examples like the hiss of loudspeakers which could be usefully considered as continuous but not differentiable.

A particularly interesting example is given by stock market prices. The reason why we know that the price of shares does not vary in a differentiable manner may be stated quite simply. If the price $f(t)$ at time t were differentiable, then, at a certain scale, it would be essentially linear, that is to say, $f(t + h) = f(t) + ah + o(h)$. Let t be the present time and let $h > 0$. Knowing $f(t - h)$ and $f(t)$, we can make a good estimate of a. If $a > 0$, we buy shares now (that is to say, at time t) and sell them at time $t + h$. If $a < 0$ we sell shares now (that is to say, at time t) and buy them at time $t + h$. If things were that simple, every fool would make money on the stock exchange. But we know that one fool's gain is another fool's loss, so that not every fool can make money on the stock exchange. Thus the proposed method cannot work and f cannot be differentiable. If the reader consults the graph of a stock market index she will see that it does indeed appear to be the case that the function pictured there is not differentiable.[20]

---

[20] However, there are advanced mathematical theories which shed a great deal of light on the behaviour of share prices and provide well-paid employment for many mathematicians. Occasionally, people tell me that some of the firms who employ these mathematicians go bankrupt. I reply that, indeed, firms which employ such mathematicians *sometimes* go bust, but those which do not *invariably* do.

I said earlier that the notion of continuity is hard-wired in our make-up. The reader may feel that the same is true of differentiation. After all, the derivative $f'(t)$ is simply the 'rate of change' of f at t. However, she will look in vain for any concept resembling 'rate of change' in classical literature. What experience would an Ancient Roman have which required such an interpretation and how would it be possible to express a rate of change using Roman numerals or measure it without accurate clocks? Ingenious historians of science have found the germ of the idea of a rate of change in certain Medieval philosophers, but it was the work of Galileo on falling bodies (see Chapter 4) which brought the idea to centre stage in mechanics. The idea then percolated outwards from physics to general mathematics and then into general 'common sense' aided by the invention of objects like the railway train which provided society with concrete illustrations of the abstract concept.[21]

Even if we are not genetically programmed with the notion of 'rate of change', the everyday life of our reader (who, if she has read this far, is certainly mathematically inclined) is so full of velocities, accelerations, rates of inflation and so on, that she will have a well-developed intuition about rates of change which I shall rely on in the work that follows. As with continuity, this means that more advanced work will require the theory to be rebuilt from scratch, but knowing the outline given here will make it easier for her to understand what the rebuilt theory will look like.

1.5 Tangents

We have already seen two ways of looking at differentiation. One was the idea of 'approximation to first order' and the second that of 'short term prediction'. There is a third way of looking at things which draws on our geometric intuition.

Suppose that we have a nice curve described, in the usual way, by $y = f(x)$. If (x_0, y_0) is a point on the curve (that is to say, if $y_0 = f(x_0)$), we can ask 'which straight line looks most like the curve near (x_0, y_0)?'. We recall that the equation for a straight line is $y = a + mx$, where m is called the *slope* or *gradient* of the line.

If we want our straight line to look 'like the curve near (x_0, y_0)', then, presumably, we want it to pass through (x_0, y_0), that is to say, we want $y_0 = a + mx_0$. If this is the case, a little algebra reveals that the equation of the line

---

[21] A cynic might add that reading the daily paper shows that many, apparently educated, individuals still misunderstand the notion.

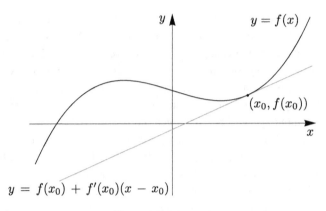

Figure 1.1 A tangent

can be rewritten as

$$y - y_0 = m(x - x_0)$$

or, after rearrangement,

$$y = f(x_0) + m(x - x_0).$$

Thus we want $f(x_0) + m(x - x_0)$ to look as close as possible to $f(x)$ when x is close to x_0. In other words (as the reader is probably already muttering to herself), we want

$$f(x_0 + \delta x) \approx f(x_0) + m\delta x$$

as closely as possible. Replacing the words 'as closely as possible' by 'to first order', we get $m = f'(x_0)$. The straight line

$$y = f(x_0) + f'(x_0)(x - x_0)$$

is called the *tangent* to the curve $y = f(x)$ at x_0.

In a work written more than two thousand years ago, Diocles recalls how 'when Zenodorus the astronomer . . . was introduced to me, he asked me how to find a mirror surface such that when it is placed facing the sun the rays reflected from it meet a point and thus cause burning.'

Diocles was able to show that a paraboloid has this property and we shall use the ideas of the calculus to provide another demonstration. However, we shall need to think quite hard and do some fairly involved calculations.[22]

---

[22] Ptolemy, King of Egypt, asked Euclid to teach him geometry. 'O King' replied Euclid 'in Egypt there are royal roads and roads for the common people, but there are no royal roads in geometry.'

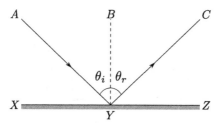

Figure 1.2 Reflection of a ray of light from a straight line

It is very easy to produce very thin (that is to say line-like) beams of light which we refer to as rays. These rays travel in straight lines and it is not hard to produce an experimental set up in which they stay within a particular plane, so we may talk about things in two rather than three dimensions. Experiment shows that, if a ray of light strikes a straight line mirror, then it is reflected in such a way that the angle[23] of incidence θ_i equals the angle of reflection θ_r as shown in Figure 1.2. (Note that the line BY is perpendicular to the mirror XZ.)

Exercise 1.5.1. *We use the notation of Figure 1.2. Explain why*

$$\tan \angle AYZ = -\tan \angle CYZ$$

so (if XZ is along the x-axis) the slope of AY is minus *the slope of CY. (This will be important when we derive the slope n of the reflected ray in a couple of paragraphs.)*

If the mirror is curved, then it is natural to expect that a ray of light hitting the mirror will be reflected as though it hit the tangent at that point, so that the angle θ_i that the incident ray makes with the perpendicular to the tangent equals the angle θ_r that the reflected ray makes with the perpendicular to the tangent as shown in Figure 1.3. Experiment confirms that this is indeed the case.

Now consider a parabolic reflector given by the equation

$$y = \frac{x^2}{4a}$$

with $a > 0$. We consider a ray of light travelling along the line $x = u$ in the direction y decreasing. It hits the parabolic mirror at the point $(u, u^2/(4a))$. Writing $f(x) = x^2/(4a)$ we see that $f'(x) = 2x/(4a)$ so the tangent at $(u, u^2/(4a))$ has a slope $u/(2a)$ and makes an angle t to the horizontal (in

---

[23] It is traditional to use the Greek letter θ (theta) for angles, but no harm will come to the reader who substitutes an ordinary letter. I think it neater to pronounce theta as 'theeta', but, sooner or later, you will hear it called 'thayta'.

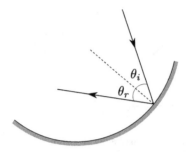

Figure 1.3 Reflection of a ray of light from a curved mirror

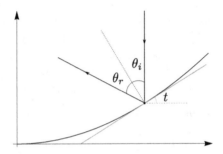

Figure 1.4 Reflection angles

the appropriate sense) given by $\tan t = u/2a$. Simple geometry and the law of reflection give us $\theta_r = \theta_i = t$ so the reflected ray makes an angle $2t$ to the vertical (in the appropriate sense) and so follows the line through $\left(u, u^2/(4a)\right)$ with slope

$$n = -\cot 2t$$

(see Figure 1.4). The doubling formula for the tangent function (see Exercise 1.5.3 (v)) gives

$$n = -\cot 2t = -\frac{1}{\tan 2t} = -\frac{1 - (\tan t)^2}{2\tan t}$$
$$= \frac{1}{2}\left(\tan t - \frac{1}{\tan t}\right) = \frac{1}{2}\left(\frac{u}{2a} - \frac{2a}{u}\right).$$

The line through $\left(u, u^2/(4a)\right)$ with slope n has the form

$$y = \frac{u^2}{4a} + n(x - u)$$

and so passes through the point $(0, u^2/(4a) - nu)$. We observe that

$$\frac{u^2}{4a} - nu = \frac{u^2}{4a} - \frac{u}{2}\left(\frac{u}{2a} - \frac{2a}{u}\right) = a$$

and so, whatever value of u we choose, the reflected ray passes through the point $(0, a)$ traditionally called the *focus* of the parabola. We have shown that all rays of light parallel to the axis[24] of a parabola are reflected so as to pass through the focus. If we move back into three dimensions and rotate the parabola about its axis we obtain a 'parabolic mirror' and we have shown that all rays of light parallel to the axis of a parabolic mirror pass through the focus

If we place a source of light at the focus, then, reversing our argument, we see that the reflected light produces a beam parallel to the axis of our mirror. This explains the use of parabolic mirrors for car headlights. Although the stars are enormous bodies, they are so distant that the light from them may be treated as a parallel beam (note another use of calculus ideas) and reflecting telescopes use parabolic mirrors to collect their light.

Exercise 1.5.2. *No one understands or believes a complicated calculation unless they have done it themselves. Close the book and try to obtain the result of this section. (Do not be disturbed if you cannot. Just re-read the appropriate part and try again.)*

Exercise 1.5.3. *This exercise is intended to remind the reader of various trigonometric formulae. You may assume the results*

$$\sin(x + y) = \sin x \cos y + \sin y \cos x$$

$$\sin(x + r) = \cos x, \quad \cos(x + r) = -\sin x, \quad \sin(-x) = -\sin x, \quad \sin r = 1,$$

where r is a right angle, but all other results must be deduced *from them.*

(i) Show that $\sin(x + 4r) = \sin x$, $\cos(x + 4r) = \cos x$, $\sin 0 = 0$ *and* $\cos(-x) = \cos x$.

(ii) By setting $x = u + r$, $y = v$ in the addition formula for sin, *or otherwise, show that*

$$\cos(u + v) = \cos u \cos v - \sin u \sin v.$$

(iii) By taking $u = -v$ in (ii), or otherwise, show that

$$(\cos u)^2 + (\sin u)^2 = 1.$$

---

[24] That is to say, the axis of symmetry.

(iv) Show that $\sin 2x = 2 \sin x \cos x$ *and*

$$\cos 2x = (\cos x)^2 - (\sin x)^2 = 1 - 2(\sin x)^2 = 2(\cos x)^2 - 1.$$

(v) Recall that $\tan x$ *is defined by* $\tan x = \sin x / \cos x$ *when* $\cos x \neq 0$. *If* $\cos x$, $\cos y$ *and* $\cos(x + y)$ *are all non-zero, show that*

$$\tan(x + y) = \frac{\tan x + \tan y}{1 - \tan x \tan y}.$$

Deduce that, if $\cos x$ *and* $\cos 2x$ *are non-zero,*

$$\tan(2x) = \frac{2 \tan x}{1 - (\tan x)^2}.$$

2

The integral

2.1 Area

Faced with a new mathematical object, we often ask 'how is it defined?' and 'how do we calculate it?'. What happens when we ask these questions about area?

It is easy to answer the second question. To find the area A of a figure, we take a sheet of graph paper divided into little squares of side s and carefully trace A onto the paper. We then count the number N of squares which lie entirely within A and the number of squares M which contain some portion of A. Then

$$Ms^2 \geq \text{area } A \geq Ns^2$$

and the difference $Ms^2 - Ns^2$ can be made as small as we please by taking s sufficiently small.

The first question is harder and many thoughtful mathematicians have chosen to answer, in effect, that the area of A is what we measure by the procedure of the previous paragraph. This answer raises the unpleasant possibility that there might be figures that do not have area either because placing A in a different way on the graph paper and choosing different values of s would produce incompatible estimates or because there do not exist choices of s which make $Ms^2 - Ns^2$ arbitrarily small.

However, it is possible, by careful book-keeping, to show that, if we use the definition just given, all polygons have area and that, if we divide a polygon into several smaller polygons, the sum of the areas of the smaller polygons equals the area of the original polygon. We shall mutter our usual incantation 'well behaved' and assume both that every well behaved region has area and that, if we divide a well behaved region into several smaller well behaved regions, the sum of the areas of the smaller well behaved regions equals the area of the original region.

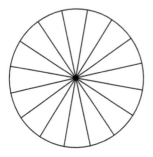 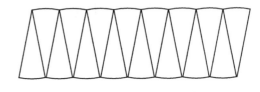

Figure 2.1 Slices of cake

I cannot resist deviating slightly from the main line of our argument to give a very informal geometrical treatment of the well-known constant π. (The ideas go back to the Ancient Greeks. Archimedes gave a full and rigorous exposition of the matter.) Let D be a disc of radius 1 (that is to say, the figure bounded by a circle of radius 1). Suppose that $u > 0$. By definition, if we place D on the graph paper considered above with s sufficiently small, the number N of squares which lie entirely within D and the number of squares M which contain some portion of D will satisfy

$$Ms^2 \geq \text{area } D \geq Ns^2 \text{ and } u \geq Ms^2 - Ns^2 \geq 0. \qquad \bigstar$$

Now look at our system through a magnifying lens, so that the disc has radius r and the graph paper is composed of squares with sides of length rs. Since nothing else has changed,

$$Mr^2s^2 \geq \text{area magnified disc} \geq Nr^2s^2. \qquad \bigstar\bigstar$$

Multiplying the relations in $\bigstar$ by r^2, we have

$$Mr^2s^2 \geq r^2 \text{ area } D \geq Nr^2s^2 \text{ and } ur^2 \geq Mr^2s^2 - Nr^2s^2 \geq 0. \qquad \bigstar\bigstar\bigstar$$

Combining $\bigstar\bigstar$ and $\bigstar\bigstar\bigstar$, we get

$$|\text{area magnified disc} - r^2 \text{ area } D| \leq ur^2.$$

Since u can be chosen as small as we want,

$$\text{area magnified disc} = r^2 \text{ area } D.$$

If we write π for the area of D, we see that the area of a disc of radius r is πr^2.

We can do more. If we take our disc of radius r and cut it up into $2n$ equal sectors, then we can rearrange them as shown in Figure 2.1 to form a figure which looks very much like a rectangle R with one side of length r and the

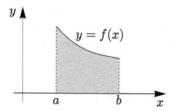

Figure 2.2 The area under the curve

adjacent side of length half the length of the perimeter of the disc. The area of R is the area of the disc, so

$$r \times \text{length perimeter disc}/2 \approx \text{product of lengths of adjacent sides of } R$$
$$= \text{area } R \approx \text{area disc} = \pi r^2$$

and, assuming that the approximation gets better and better as n increases,

$$r \times \text{length of the perimeter disc}/2 = \pi r^2,$$

so the length of the perimeter of a circle of radius r is $2\pi r$.

2.2 Integration

If f is a well behaved function, $a \leq b$ and $0 \leq f(t)$ for $a \leq t \leq b$, then we denote 'the area under the curve $y = f(x)$' by $\int_a^b f(t)\,dt$. More precisely, we say that the area of the figure $A(f)$ bounded by the three straight line segments consisting of the points (a, y) with $0 \leq y \leq f(a)$; the points (b, y) with $0 \leq y \leq f(b)$; the points $(x, 0)$ for $a \leq x \leq b$; and the curve $(x, f(x))$ with $a \leq x \leq b$ is

$$\int_a^b f(t)\,dt$$

pronounced[1] 'the integral of f from a to b'. We illustrate this in Figure 2.2.[2] Note that, if $a = b$, then $\int_a^b f(t)\,dt = 0$ automatically.

---

[1] Why do we use this notation? Because it is used by every mathematician from China to Peru. (The refusal of Cambridge to adopt the standard continental notation until about 1820 was both a symptom and a cause of that university's mathematical backwardness.) Where does it come from? Leibniz wished to exploit the analogy between integrals and sums and thought of $\int$ as an elongated s as in the Latin 'summa'. The placing of a and b as subscript and superscript is due to Fourier.

[2] The reader may prefer to think of $A(f)$ as the collection of points (x, y) with $0 \leq y \leq f(x)$ and $a \leq x \leq b$.

The reader is entitled to ask 'what do we know about the integral which the Greeks did not know about area?'. I answer that we know the following key result. If $f(t)$ and $g(t)$ are well behaved functions of t, and $0 \le f(t)$, $g(t)$ for $a \le t \le b$, then

$$\int_a^b f(t) + g(t)\,dt = \int_a^b f(t)\,dt + \int_a^b g(t)\,dt.$$

To show this, consider our sheet of graph paper with $s = (b - a)/n$, where n is some large integer. We choose $A(f)$ as before and, for each integer r with $1 \le r \le n$, take $A_r(f)$ to be the figure bounded by the three straight line segments consisting of the points $(a + (r - 1)s, y)$ with $0 \le y \le f(a + (r - 1)s)$; the points $(a + rs, y)$ with $0 \le y \le f(a + rs)$; the points $(x, 0)$ for $a + (r - 1)s \le x \le a + rs$; and the curve $(x, f(x))$ with $a + (r - 1)s \le x \le a + rs$. Let $N_r(f)$ be the number of squares which lie entirely within $A_r(f)$ and $M_r(f)$ be the number of squares which contain some portion of $A_r(f)$. If $N(f)$ is the number of squares which lie entirely within $A(f)$ and $M(f)$ is the number of squares which contain some portion of $A(f)$, we have

$$N(f) = N_1(f) + N_2(f) + \cdots + N_n(f),$$
$$M(f) = M_1(f) + M_2(f) + \cdots + M_n(f).$$

Similarly,

$$N(g) = N_1(g) + N_2(g) + \cdots + N_n(g),$$
$$M(g) = M_1(g) + M_2(g) + \cdots + M_n(g)$$

and

$$N(f + g) = N_1(f + g) + N_2(f + g) + \cdots + N_n(f + g),$$
$$M(f + g) = M_1(f + g) + M_2(f + g) + \cdots + M_n(f + g).$$

We now observe that

$$f(t) \le M_r(f)s \text{ and } g(t) \le M_r(g)s$$

so

$$f(t) + g(t) \le \big(M_r(f) + M_r(g)\big)s$$

whenever $a + (r - 1)s \le t \le a + rs$, whence

$$M_r(f + g) \le M_r(f) + M_r(g).$$

Exercise 2.2.1. *Show, in the same way, that*

$$N_r(f) + N_r(g) \le N_r(f + g).$$

Combining these results with those of the previous paragraph, we get

$$N(f) + N(g) \le N(f + g) \le M(f + g) \le M(f) + M(g).$$

Thus

$$N(f)s^2 + N(g)s^2 \le N(f + g)s^2 \le A(f + g) \le M(f + g)s^2$$
$$\le M(f)s^2 + M(g)s^2.$$

But

$$N(f)s^2 \le A(f) \le M(f)s^2, \quad M(g)s^2 \le A(g) \le N(g)s^2$$

and we can make $M(f)s^2 - N(f)s^2$ and $M(g)s^2 - N(g)s^2$ as small as we wish by taking s sufficiently small, so

$$A(f + g) = A(f) + A(g).$$

In other words,

$$\int_a^b f(t) + g(t)\, dt = \int_a^b f(t)\, dt + \int_a^b g(t)\, dt.$$

Exercise 2.2.2. *Restate the algebraic argument that we have just given in a geometric form involving sliding blocks around in the same way that we slid bits of cake to produce Figure 2.1 on page 30.*

The result just proved is remarkable in itself, but it also enables us to extend the notion of an integral to functions f which are not necessarily positive.[3] The idea is to write f as the difference of two positive functions g and h and define the integral of f as the difference of the integrals of g and h. However, we can choose g and h in many different ways, so we must show that *different* choices lead to the *same* result.

Exercise 2.2.3. *Let f be a function.*

(i) Show that, if we set

$$g(t) = \begin{cases} f(t) & \text{if } 0 \le f(t), \\ 0 & \text{otherwise} \end{cases}$$

---

[3] It is very easy to wave one's hands and talk about 'negative areas'. What we have to do is to show that we can combine the ideas of 'negative' and 'positive' areas in a consistent manner.

and $h(t) = g(t) - f(t)$, *then* $f(t) = g(t) - h(t)$ *and* $0 \le g(t)$, $h(t)$ *whenever* $a \le t \le b$.

(ii) *Show that, if we set*

$$g(t) = |f(t)| \text{ and } h(t) = |f(t)| - f(t),$$

then $f(t) = g(t) - h(t)$ *and* $0 \le g(t)$, $h(t)$ *whenever* $a \le t \le b$.

(iii) *Suppose that* $a \le b$, $0 \le M$ *and* $-M \le f(t)$ *whenever* $a \le t \le b$. *Show that, if we set*

$$g(t) = M + f(t) \text{ and } h(t) = M,$$

then $f(t) = g(t) - h(t)$ *and* $0 \le g(t)$, $h(t)$ *whenever* $a \le t \le b$.

The freedom to choose g and h in different ways is very useful in the arguments that follow.

How do we show that our extended definition of the integral is unambiguous? Observe that, if g_1, g_2, h_1 and h_2 are well behaved functions, $g_1(t)$, $g_2(t)$, $h_1(t)$, $h_2(t) \ge 0$ and

$$g_1(t) - h_1(t) = g_2(t) - h_2(t)$$

for $a \le t \le b$, then

$$g_1(t) + h_2(t) = g_2(t) + h_1(t)$$

so that

$$\int_a^b g_1(t)\,dt + \int_a^b h_2(t)\,dt = \int_a^b g_1(t) + h_2(t)\,dt$$

$$= \int_a^b g_2(t) + h_1(t)\,dt$$

$$= \int_a^b g_2(t)\,dt + \int_a^b h_1(t)\,dt$$

and so

$$\int_a^b g_1(t)\,dt - \int_a^b h_1(t)\,dt = \int_a^b g_2(t)\,dt - \int_a^b h_2(t)\,dt.$$

Thus we can make the following *unambiguous* definition. If $f(t) = g(t) - h(t)$, where $g(t)$ and $h(t)$ are well behaved functions of t, and $0 \le g(t)$, $h(t)$, then we set

$$I_a^b f(t)\,dt = \int_a^b g(t)\,dt - \int_a^b h(t)\,dt. \qquad \bigstar$$

Exercise 2.2.4. *Suppose that $F(t)$ is a well behaved function with $0 \le F(t)$ for $a \le t \le b$. By considering $G(t) = F(t)$, $H(t) = 0$ show that*

$$I_a^b F(t)\,dt = \int_a^b F(t)\,dt.$$

The exercise suggests that we should write

$$I_a^b f(t)\,dt = \int_a^b f(t)\,dt$$

and we shall do so. Equation ★ then takes the form

$$\int_a^b f(t)\,dt = \int_a^b g(t)\,dt - \int_a^b h(t)\,dt. \qquad\qquad ★★$$

Exercise 2.2.5. *If $a \le b$ and f is a well behaved function, show that*

$$\int_a^b \big(-f(t)\big)\,dt = -\int_a^b f(t)\,dt.$$

If we want to demonstrate a result about $\int_a^b f(t)\,dt$, one natural method is first to demonstrate it for the case when $0 \le f(t)$ for $a \le t \le b$ and then use equation ★★ to extend the result to the case when $f(t)$ is not necessarily positive. The reader should use this method in the next exercise.

Exercise 2.2.6. *If $a \le b$ and f and g are well behaved functions, show, by taking $f(t) = f_1(t) - f_2(t)$, $g(t) = g_1(t) - g_2(t)$ and using the result for functions taking only positive values, that*

$$\int_a^b f(t) + g(t)\,dt = \int_a^b f(t)\,dt + \int_a^b g(t)\,dt.$$

Show also that

$$\int_a^b f(t) - g(t)\,dt = \int_a^b f(t)\,dt - \int_a^b g(t)\,dt.$$

We deduce a simple but very important result which we call the *inequality rule for integration*. Suppose that f and g are well behaved functions, $a \le b$ and $g(t) \le f(t)$ whenever $a \le t \le b$. Then $0 \le f(t) - g(t)$ whenever $a \le t \le b$ so

$$0 \le \int_a^b f(t) - g(t)\,dt = \int_a^b f(t)\,dt - \int_a^b g(t)\,dt$$

and

$$\int_a^b g(t)\,dt \le \int_a^b f(t)\,dt.$$

We shall make use of this result over and over again in the rest of this book.

Exercise 2.2.7. *(i) Suppose that f is a well behaved function, $a \leq b$ and $0 \leq f(t)$ for $a \leq t \leq b$. If p is a strictly positive integer, use the fact that*

$$pf(t) = \underbrace{f(t) + f(t) + \cdots + f(t)}_{p}$$

to show that

$$\int_a^b pf(t)\,dt = p\int_a^b f(t)\,dt.$$

Why is the result also true when $p = 0$?

(ii) Let f be as in (i). By considering

$$q\int_a^b \frac{p}{q}f(t)\,dt,$$

show that

$$\int_a^b \frac{p}{q}f(t)\,dt = \frac{p}{q}\int_a^b f(t)\,dt$$

whenever p and q are integers with $0 \leq p$, $1 \leq q$.

(iii) Let f be as in (i) and let $0 \leq u$. If q is an integer with $1 \leq q$, let p be the integer such that

$$\frac{p}{q} \leq u < \frac{p+1}{q}.$$

Show, using the inequality rule for integrals, that

$$\frac{p}{q}\int_a^b f(t)\,dt \leq \int_a^b uf(t)\,dt \leq \frac{p+1}{q}\int_a^b f(t)\,dt.$$

Use the fact that we can take q as large as we wish to deduce that

$$\int_a^b uf(t)\,dt = u\int_a^b f(t)\,dt.$$

(iv) Deduce that, if g is a well behaved (but not necessarily positive) function and v a constant (which need not be positive), then

$$\int_a^b vg(t)\,dt = v\int_a^b g(t)\,dt.$$

We conclude this section by making a further extension to the notion of an integral. It may strike the reader as both odd and unnecessary, but experience will show that it is rather useful. If f is a well behaved function and $a \leq b$, we write

$$\int_b^a f(t)\,dt = -\int_a^b f(t)\,dt.$$ †

Exercise 2.2.8. *Suppose that f and g are well behaved functions. Show that, whatever the values of a and b, the following results hold.*

(i) $\displaystyle\int_a^b f(t) + g(t)\,dt = \int_a^b f(t)\,dt + \int_a^b g(t)\,dt.$

(ii) *If v is a constant, then*

$$\int_a^b vf(t)\,dt = v\int_a^b f(t)\,dt.$$

(iii) *(Both a useful remark and a warning.)*

If $\displaystyle\int_a^b g(t)\,dt \leq \int_a^b f(t)\,dt$, *then* $\displaystyle\int_b^a f(t)\,dt \leq \int_b^a g(t)\,dt.$

If $a \leq b \leq c$ and f is a well behaved function with $0 \leq f(t)$ for $a \leq t \leq c$, then the definition of the integral as the area under the curve shows that

$$\int_a^b f(t)\,dt + \int_b^c f(t)\,dt = \int_a^c f(t)\,dt.$$

Exercise 2.2.9. *If $a \leq b \leq c$ and f is a well behaved (but not necessarily positive) function, show that*

$$\int_a^b f(t)\,dt + \int_b^c f(t)\,dt = \int_a^c f(t)\,dt.$$

Exercise 2.2.10. *Suppose that f is a well behaved function.*

(i) *Suppose that $b \leq c \leq a$. Use the fact that*

$$\int_b^c f(t)\,dt + \int_c^a f(t)\,dt = \int_b^a f(t)\,dt$$

to show that

$$\int_a^b f(t)\,dt + \int_b^c f(t)\,dt = \int_a^c f(t)\,dt.$$

(ii) *Show, similarly, that, if $c \leq b \leq a$,*

$$\int_a^b f(t)\,dt + \int_b^c f(t)\,dt = \int_a^c f(t)\,dt.$$

(iii) *Show that, whatever the order of a, b and c,*

$$\int_a^b f(t)\,dt + \int_b^c f(t)\,dt = \int_a^c f(t)\,dt.$$

(iv) *Suppose that $f(t) = k$ for all t. Show that, whatever the order of a and b,*

$$\int_a^b f(t)\,dt = k(b - a).$$

This result is usually written in the form

$$\int_a^b k\,dt = k(b - a).$$

We seem, in the words of Barrow,[4] to have 'built very large gates for a very small city' since, so far, the only function for which we know the integral is the constant function! The next exercise shows that we can do slightly better.

Exercise 2.2.11. *(i) Show that $r = \frac{1}{2}\big((r + 1)r - r(r - 1)\big)$ and deduce that*

$$1 + 2 + \cdots + n = \frac{1}{2}n(n + 1).$$

(ii) *Show that $r(r - 1) = \frac{1}{3}\big((r + 1)r(r - 1) - r(r - 1)(r - 2)\big)$ and deduce that*

$$1 \times 0 + 2 \times 1 + \cdots + r(r - 1) + \cdots + n(n - 1) = \frac{1}{3}(n + 1)n(n - 1).$$

(iii) *Use the previous results to show that*

$$1^2 + 2^2 + \cdots + r^2 + \cdots + n^2 = \frac{1}{6}n(n + 1)(2n + 1).$$

(iv) *Use the inequality rule for integration (see page 35) to show that*

$$\int_{r/n}^{(r+1)/n} \frac{r^2}{n^2}\,dt \le \int_{r/n}^{(r+1)/n} t^2\,dt \le \int_{r/n}^{(r+1)/n} \frac{(r + 1)^2}{n^2}\,dt$$

and deduce that

$$\frac{r^2}{n^3} \le \int_{r/n}^{(r+1)/n} t^2\,dt \le \frac{(r + 1)^2}{n^3}.$$

(v) *By adding the inequalities of (iv), show that*

$$\frac{1}{n^3}(0^2 + 1^2 + \cdots + (n - 1)^2) \le \int_0^1 t^2\,dt \le \frac{1}{n^3}(1^2 + 2^2 + \cdots + n^2)$$

[4] *Geometrical Lectures*, Lecture VIII.

and so

$$\frac{(1-n^{-1})(1-\tfrac{1}{2}n^{-1})}{3} \le \int_0^1 t^2 \, dt \le \frac{(1+n^{-1})(1+\tfrac{1}{2}n^{-1})}{3}.$$

Conclude that

$$\int_0^1 t^2 \, dt = \frac{1}{3}.$$

(vi) Show that, if a > 0,

$$\int_{ra/n}^{(r+1)a/n} \frac{a^2}{n^2} \, dt \le \int_{ra/n}^{(r+1)a/n} t^2 \, dt \le \int_{ra/n}^{(r+1)a/n} \frac{a^2}{n^2} \, dt$$

and deduce that

$$\int_0^a t^2 \, dt = \frac{a^3}{3}.$$

Show that the result also holds for a ≤ 0 and deduce that

$$\int_a^b t^2 \, dt = \frac{b^3 - a^3}{3}$$

for all values of a and b.

Exercise 2.2.12.• *We develop the ideas of the previous exercise a little more. Take k to be a strictly positive integer.*

(i) If $u_r = r(r-1)(r-2)\cdots(r-k+1)$ show that

$$u_1 + u_2 + \cdots + u_n = \frac{1}{k+1}(n+1)n(n-1)(n-2)\cdots(n-k+1).$$

(ii) Show that

$$1^k + 2^k + \cdots + n^k = \frac{n^{k+1}}{k+1} + P_k(n),$$

where P_k is some polynomial of degree at most k which need not be specified further.

(iii) Show that

$$\int_a^b t^k \, dt = \frac{b^{k+1} - a^{k+1}}{k+1}.$$

The computation of integrals (or, as they would have viewed them, areas) of the type just given required the ingenuity of great mathematicians like Archimedes, Cavalieri, Fermat and Wallis. In the next section we shall see how a new idea enables very ordinary mortals to perform the same feats.

2.3 The fundamental theorem

As we have seen, the Ancient Greeks were familiar with both the notions of tangent and area. It was left to the mathematicians of the seventeenth century to discover the strong connection between the two which is now known as the *fundamental theorem of the calculus*.[5] The result was first written down in full generality by Barrow, one of those larger than life figures from the days before the professionalisation of mathematics.[6]

Neither Barrow, nor anyone else, seems to have attached much importance to his result. One reason may be that Barrow saw it as a theorem in geometry and its real power is only visible in the algebraic presentation of the calculus. Another may be that Barrow thought that it was not possible to do more than add a few improvements to Greek mathematics. The younger generation of Newton and Leibniz believed that it was possible to go far beyond the Ancients and that they were the men to do it.

Even when the fundamental theorem was incorporated into the splendid structure built by Newton and Leibniz, it was just considered as one result among many – partly no doubt because, as we shall see, it is very easy to demonstrate. (Indeed, in Leibniz's version of the calculus, it looks very much like a tautology.)

When mathematicians rethought the structure of the calculus in the nineteenth century, it became clear that the theorem was a very important part of the structure and it acquired its present impressive title. The twentieth century saw several extensions and generalisations of the ideas of area and of tangent. Attempts to apply the fundamental theorem in these new contexts increased the respect of mathematicians for its importance and subtlety.[7]

---

[5] Often just the 'fundamental theorem of calculus'.

[6] Barrow was successively Professor of Greek at Cambridge, the first Lucasian Professor of Mathematics at Cambridge, a leading preacher (Chaplain to the King) and Master of Trinity College. He is still remembered for his success in each of these four roles. His wit and pugnacity are shown in an exchange with Lord Rochester, a man as famous for being wicked as Barrow was for being good.

> *Rochester, bowing deeply*: Doctor, I am yours to the shoe-tie.
> *Barrow, bowing still deeper*: My lord, I am yours to the ground.
> *Rochester*: Doctor, I am yours to the centre.
> *Barrow*: My lord, I am yours to the antipodes.
> *Rochester*: Doctor, I am yours to the lowest pit of hell.
> *Barrow, turning on his heel*: There, my lord, I leave you.

[7] In view of the complex history of the most important generalisation of the fundamental theorem, Arnold suggested that it be called the Barrow–Newton–Leibniz–Gauss–Green–Ostrogradski–Kelvin–Stokes–Poincaré theorem, but it is more usually called Stokes' theorem.

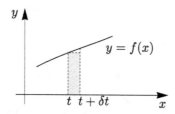

Figure 2.3 The fundamental theorem

After such an opening fanfare, the reader may be surprised to learn that, in effect, the fundamental theorem is simply the observation that, if f is a well behaved function, then

$$\int_t^{t+\delta t} f(x)\,dx = f(t)\delta t + o(\delta t). \qquad \bigstar$$

If $0 < \delta t$, and $f(s)$ is positive for $t \le s \le t + \delta t$, this is nothing but the statement that the area of the figure bounded by the three straight line segments consisting of the occasions (t, y) with $0 \le y \le f(t)$; the points $(t + \delta t, y)$ with $0 \le y \le f(t + \delta t)$; the points $(x, 0)$ with $t \le x \le t + \delta t$ and the curve $(x, f(x))$ with $t \le x \le t + \delta t$ is well approximated by the area of the rectangle with vertices $(t, 0)$, $\big(t, f(t)\big)$, $\big(t + \delta t, f(t)\big)$ and $(t + \delta t, 0)$ when δt is small. We illustrate this in Figure 2.3.

Exercise 2.3.1. *(i) Explain the case when $\delta t < 0$ and $f(s)$ is positive for $t + \delta t \le s \le t$ by drawing a similar diagram.*
(ii) Use the fact that

$$\int_a^b -g(t)\,dt = -\int_a^b g(t)\,dt$$

to obtain the result when $\delta t > 0$ and and $f(s)$ is negative for $t \le s \le t + \delta t$.

The reader should not be satisfied until she considers equation $\bigstar$ to be *obvious*. However, in view of the importance of this result, we give a more formal treatment. Since f is continuous, Definition 1.3.3 tells us that, given any $u > 0$, we can find a $v > 0$ such that $|f(t + h) - f(t)| \le u$ whenever $|h| \le v$. In other words, we have

$$f(t) - u \le f(x) \le f(t) + u$$

whenever $|x - t| \le v$. Thus, if $0 \le \delta t \le v$, we have

$$\int_t^{t+\delta t} (f(t) - u)\,dx \le \int_t^{t+\delta t} f(x)\,dx \le \int_t^{t+\delta t} (f(t) + u)\,dx$$

so that (noting that $f(t) - u$ and $f(t) + u$ are constants)

$$(f(t) - u)\delta t \leq \int_t^{t+\delta t} f(x)\,dx \leq (f(t) + u)\delta t.$$

Exercise 2.3.2. *Show that, if $0 \leq \delta t \leq v$, we have*

$$(f(t) - u)(-\delta t) \geq \int_t^{t-\delta t} f(x)\,dx \geq (f(t) + u)(-\delta t).$$

[*Notice that this is one of the points where it is useful to have the extension of the notion of integration to cover integrals from a to b when b < a. See the discussion starting with formula † on page 37.*]

Thus, whatever the sign of δt,

$$\left| \int_t^{t+\delta t} f(x)\,dx - f(t)\delta t \right| \leq u|\delta t|$$

whenever $|\delta t| \leq v$ and so ★ holds.

If we write

$$F(t) = \int_a^t f(x)\,dx,$$

then equation ★ tells us that

$$F(t + \delta t) = \int_a^{t+\delta t} f(x)\,dx = \int_a^t f(x)\,dx + \int_t^{t+\delta t} f(x)\,dx$$

$$= F(t) + \int_t^{t+\delta t} f(x)\,dx = F(t) + f(t)\delta t + o(\delta t).$$

In other words, $F(t)$ is a differentiable function of t with $F'(t) = f(t)$. We have shown that differentiation reverses the effect of integration.

Is it true that integration reverses the effect of differentiation? The following example shows that things are not *quite* this simple. If $G(t) = 1$ for all t, then

$$\int_a^t G'(x)\,dx = \int_a^t 0\,dx = 0 \neq G(t).$$

However, things are *almost* as simple as we could desire. Suppose that $F(t)$ is differentiable and its derivative $f(t) = F'(t)$ is continuous. Then, if we set

$$g(t) = F(t) - F(a) - \int_a^t F'(x)\,dx,$$

the result we have just proved tells us that

$$g'(t) = F'(t) - 0 - F'(t) = 0$$

for all t. Since the rate of change $g'(t) = 0$ everywhere, it is surely obvious[8] that $g(t)$ must be constant. Since

$$g(a) = F(a) - F(a) - \int_a^a F'(x)\,dx = F(a) - F(a) - 0 = 0,$$

we must have $g(t) = 0$ for all t and so

$$F(t) = F(a) + \int_a^t F'(x)\,dx.$$

We write

$$F(t) - F(a) = [F(x)]_a^t$$

and obtain the standard formula

$$\int_a^t F'(x)\,dx = [F(x)]_a^t.$$

We thus have the following procedure for evaluating an integral $\int_a^b f(x)\,dx$.

(1) *Guess* a function $F(x)$ such that $F'(x) = f(x)$.

(2) *Check* that your guess is correct.

(3) We now *know* that

$$\int_a^b f(x)\,dx = [F(x)]_a^b = F(b) - F(a).$$

We shall see in Section 3.1 that there may be no simple expression for F and, of course, even if there is a simple expression, we may not be clever enough to find it.

When we evaluate

$$\int_a^b f(t)\,dt$$

we say that 'we integrate f from a to b'. Many people (but not the author) call $\int_a^b f(t)\,dt$ the definite integral and $F(t) + c$ with c an arbitrary constant the indefinite integral.[9]

Exercise 2.3.3. *(Not very different from what we have already done.) Suppose that $F'(x) = G'(x) = f(x)$. By differentiating g defined by*

$$g(x) = \big(F(x) - G(x)\big) - \big(F(a) - G(a)\big),$$

show that there is a constant c such that $G(x) = F(x) + c$ for all x.

---

[8] But note that I provide no proof. The suspicious reader must wait until the final chapter for a fuller discussion.

[9] Other conventions exist, but those who use this particular convention make a shibboleth out of the words 'plus an arbitrary constant'.

Suppose, conversely, that $F'(x) = f(x)$ and $G(x) = F(x) + c$ for some constant c. Show that $G'(x) = f(x)$.
[*The contents of this exercise are usually expressed by a statement along the lines of 'the anti-derivative is unique up to the addition of a constant'.*]

Exercise 2.3.4. (*i*) *Compute*

$$\int_0^x (a_0 + a_1 t + a_2 t^2 + \ldots + a_n t^n)\, dt.$$

Of course, we could obtain the result of this question by using Exercise 2.2.12 and various integral formulae, but the reader should observe how much easier it is to use the fundamental theorem.
   (*ii*) *Let p and q be integers with $q \neq 0$ and $p \neq -q$. If $b > a > 0$, find*

$$\int_a^b t^{p/q}\, dt.$$

It is much less evident how we might obtain this result without using the fundamental theorem. Fermat used a very clever argument to do this, but, as we have done, left open the problem of finding $\int_a^b t^{-1}\, dt$ This open case was known as the 'problem of the quadrature of the hyperbola'. It will do the reader no harm to spend a few minutes trying to use the tools we have developed so far to solve it. We will discuss the problem further in Section 3.1 and use the results of our discussion throughout the rest of the book.

We shall not evaluate many integrals in the course of this book, but we note that the various rules for differentiation are reflected in useful tricks for integration. In what follows we assume that all our functions are well behaved.

Integration by parts. Suppose that $F'(x) = f(x)$ and $G'(x) = g(x)$. By the product rule (see page 19) applied to $H(x) = F(x)G(x)$ we have

$$[F(x)G(x)]_a^b = [H(x)]_a^b = \int_a^b H'(x)\, dx$$

$$= \int_a^b \left(F'(x)G(x) + F(x)G'(x) \right) dx$$

$$= \int_a^b F'(x)G(x)\, dx + \int_a^b F(x)G'(x)\, dx$$

$$= \int_a^b f(x)G(x)\, dx + \int_a^b F(x)g(x)\, dx.$$

Rewriting, we obtain the standard rule for 'integration by parts'

$$\int_a^b f(x)G(x)\,dx = [F(x)G(x)]_a^b - \int_a^b F(x)g(x)\,dx.$$

Change of variable. Suppose that $F'(x) = f(x)$ and $G'(x) = g(x)$. By the function of a function rule (see page 19) applied to $H(x) = F\big(G(x)\big)$ we have

$$\int_{G(a)}^{G(b)} f(y)\,dy = [F(y)]_{G(a)}^{G(b)} = [H(x)]_a^b$$

$$= \int_a^b H'(x)\,dx = \int_a^b G'(x)F'\big(G(x)\big)\,dx$$

$$= \int_a^b g(x)f\big(G(x)\big)\,dx.$$

There is great deal more that can said about the 'change of variable formula', but a first glance at calculus cannot cover everything.

Exercise 2.3.5. *We can now turn the ideas of Exercise 2.2.11 on their head. Suppose that $a = p/q$ with p and q strictly positive integers.*

(i) Show that

$$\frac{1^a + 2^a + \cdots + (n-1)^a}{n^{a+1}} \le \int_0^1 t^a\,dt \le \frac{1^a + 2^a + \cdots + n^a}{n^{a+1}}.$$

(ii) Use Exercise 1.4.7 to evaluate the integral and deduce that

$$\frac{1^a + 2^a + \cdots + (n-1)^a}{n^{a+1}} \le \frac{1}{a+1} \le \frac{1^a + 2^a + \cdots + n^a}{n^{a+1}}.$$

(iii) Hence show that

$$0 \le \frac{1^a + 2^a + \cdots + n^a}{n^{a+1}} - \frac{1}{a+1} \le \frac{1}{n}.$$

2.4 Growth

Throughout this book we deal with the old-style calculus with its emphasis on common sense and good behaviour. Most of what we say passes over to the new-style analysis that the reader will meet in more advanced texts with minor, though important, changes. I should, however, warn the reader that, though the results of this particular section remain correct in the new analysis, the part that they play in the general argument is very different.[10]

---

[10] In particular, in some places where we obtain *B* from *A*, the new analysis proves *A* from *B*.

Suppose that f' is continuous and that $f'(t) \geq 0$ for $a \leq t \leq b$. Then, by the fundamental theorem of the calculus,

$$f(b) - f(a) = \int_a^b f'(t)\,dt \geq 0.$$

More briefly, a well behaved function with positive derivative must be increasing.

In the other direction, we observe that if

$$f(t + \delta t) = f(t) + A\delta t + o(\delta t)$$

and $A < 0$, then we will have $f(t + \delta t) < f(t)$ for δt small and positive. Thus, if $f(s)$ is increasing for $a \leq s \leq b$, we must have $f'(s) \geq 0$ for for $a < s < b$. More briefly, a well behaved increasing function has positive derivative.

Exercise 2.4.1. *If $f(s) = s^3$, show that $f(s)$ is strictly increasing (that is to say, $f(a) < f(b)$ whenever $a < b$),[11] but $f'(0) = 0$. (Thus a strictly increasing function need not have a strictly positive derivative.)*

These results are linked to a result that we call the *mean value inequality*. If f is a well behaved function with $|f'(t)| \leq M$ for $a \leq t \leq b$, then

$$|f(a) - f(b)| \leq M|b - a|.$$

Exercise 2.4.2 shows that we can derive this inequality from the results already obtained, but we shall derive it directly as follows. The result for $b \leq a$ follows from the result for $a \leq b$, so we may suppose that $a \leq b$. Since

$$-M \leq f'(t) \leq M,$$

we have

$$\int_a^b (-M)\,dt \leq \int_a^b f'(t)\,dt \leq \int_a^b M\,dt$$

and so

$$-M(b - a) \leq f(b) - f(a) \leq M(b - a).$$

Exercise 2.4.2. *(An alternative demonstration.) Suppose that f is well behaved, $a \leq b$ and $|f'(t)| \leq M$ for $a \leq t \leq b$. If we set $g(t) = f(t) - Mt$, show that $g'(t) \leq 0$ for $a \leq t \leq b$ and deduce that $g(b) \leq g(a)$. Conclude that $f(b) - f(a) \leq M(b - a)$. Show similarly that $-M(b - a) \leq f(b) - f(a)$.*

---

[11] In the same spirit, mathematicians say that b is *strictly greater* than b when $b > a$ and a is *strictly positive* when $a > 0$.

One way of expressing the mean value inequality is to say that, if the rate of change of a function of time is never greater than M, then it cannot change by more than MT over a time T.

The mean value inequality looks unimpressive, but turns out to be very useful in all sorts of circumstances. We give a development of this inequality when we discuss Taylor's theorem with remainder estimate in Section 6.2.

Exercise 2.4.3. *Let $a < b$. We have shown that, if f is a well behaved function with $f'(x) \geq 0$ whenever $a \leq x \leq b$, then $f(t) \leq f(s)$ whenever $a \leq t \leq s \leq b$. (That is to say, f is increasing.)*

If $f'(x) > 0$ whenever $a \leq x \leq b$, we can make the following improvement. Suppose that $a \leq t < s \leq b$. Set $y = (t + s)/2$. Explain why we can find a $u > 0$ with $u \leq (t - s)/2$ such that

$$|f(y + h) - f(y) - f'(y)h| \leq \frac{f'(y)}{2}|h|$$

whenever $|h| \leq u$. Show that

$$f(y + u) \geq f(y) + \frac{f'(y)}{2}u$$

and deduce that $f(t) < f(s)$. (That is to say, f is strictly increasing.)

2.5 Maxima and minima

Consider the following problem.

Problem A. *The river Camazon runs between parallel banks $y = 0$ and $y = b$, where $b > 0$. Supergirl is bird watching at $(0, 0)$ on one bank when she observes Superman being attacked by a crocodile at (a, b) on the other bank. (We take $a \geq 0$.) Supergirl instantly decides to run along the bank to $(s, 0)$ and then dive in and swim directly to (a, b). If Supergirl can run at speed u and swim at speed v, what choice of s will get her to (a, b) as quickly as possible?*

Exercise 2.5.1. *(i) Explain why we need only consider the case $s \geq 0$.*
(ii) Show that the time $f(s)$ taken is given by

$$f(s) = \frac{s}{u} + \frac{\sqrt{(a - s)^2 + b^2}}{v}.$$

The discussion of the previous section suggests that we try using differentiation to attack our problem. We shall be led in a natural manner to consider, not only the derivative f' of f, but also the derivative of f' itself. We write

$f''(t) = (f')'(t)$ for the derivative of the derivative and call it the *second derivative* of f. We pronounce f'' as 'f double dash'.

Exercise 2.5.2. *(i) Use the rules for differentiation to show that*

$$f'(s) = \frac{1}{u} - \frac{1}{v} \frac{a - s}{\left((a - s)^2 + b^2\right)^{1/2}}.$$

(ii) Show further that

$$f''(s) = \frac{1}{v} \left(\frac{1}{\left((a - s)^2 + b^2\right)^{1/2}} - \frac{(a - s)^2}{\left((a - s)^2 + b^2\right)^{3/2}} \right)$$

$$= \frac{1}{v} \frac{b^2}{\left((a - s)^2 + b^2\right)^{3/2}} > 0.$$

By Exercises 2.5.2 (ii) and 2.4.3, $f'(s)$ increases strictly as s increases. We also see from (i) that $f'(s) \geq 1/u > 0$ when $s \geq a$.

There are two possibilities.

(1) If

$$f'(0) = \frac{1}{u} - \frac{1}{v} \frac{a}{(a^2 + b^2)^{1/2}} \geq 0,$$

then $f'(s) > f'(0) = 0$ whenever $s > 0$, so $f(s)$ is increasing for $s \geq 0$ and attains its minimum when $s = 0$. Supergirl should dive in at once.

(2) If

$$f'(0) = \frac{1}{u} - \frac{1}{v} \frac{a}{(a^2 + b^2)^{1/2}} < 0,$$

then (since $f'(a) > 0$) there will be an s_0 with $0 \leq s_0 \leq a$ and $f'(s_0) = 0$, that is to say, with

$$\frac{v}{u} = \frac{a - s_0}{\left((a - s_0)^2 + b^2\right)^{1/2}}.$$

We will have $f'(s) > 0$ whenever $s > s_0$, so $f(s)$ is increasing in the range $s > s_0$ and we will have $f'(s) < 0$ whenever $s < s_0$, so $f(s)$ is decreasing in the range $s < s_0$. Thus $f(s)$ attains its minimum when $s = s_0$ and Supergirl should run to $(s_0, 0)$ and then dive in.

The next exercise gives a nice geometric interpretation of this result.

Exercise 2.5.3. *(i) If $v \geq u$, show from our results that Supergirl should always dive in at once. Why is this obvious directly?*

(ii) Now suppose that $u \geq v$ and $\sin \theta_0 = v/u$. Show that, if the line through $(0, 0)$ and (a, b) makes an angle less than θ_0 with the y-axis, Supergirl should

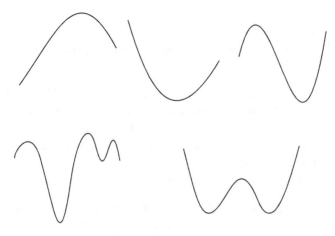

Figure 2.4 A child's garden of maxima

dive in at once, but that, otherwise, she should run the point $(s_0, 0)$ such that the line through $(s_0, 0)$ and (a, b) makes an angle θ_0 with the y-axis and dive in there.

(iii) Is part (ii) obvious apart from the value of θ_0?

Let us try and put the method of the previous exercise into a more general context. First we note that, as a look at Figure 2.4 shows, even the question 'for what values of t with $a \leq t \leq b$ does the well behaved function $f(t)$ of t attain its largest value?' is not as simple as it looks.

Exercise 2.5.4. *Sketch the curves given in Figure 2.4 and mark the points where the maximum occurs.*

Inspection of these examples suggests that it may be useful to distinguish between 'global' and 'local' maxima. Suppose that we are interested in the values of a well behaved function $f(t)$ for $a \leq t \leq b$. We say that f has a *global maximum* at s if $a \leq s \leq b$ and $f(t) \leq f(s)$ whenever $a \leq t \leq b$. We say that f has a *local maximum* at s with $a \leq s \leq b$ if we can find a $u > 0$ such that whenever $a \leq t \leq b$ and $|t - s| \leq u$ we have $f(s) \geq f(t)$. The easy exercise that follows simply puts the definition in different words.

Exercise 2.5.5. *Show that our definition is equivalent to saying that f has a local maximum at s with $a \leq s \leq b$ if one of the following conditions hold:*

(1) $a < s < t$ and we can find a $u > 0$ such that $f(t) \leq f(s)$ whenever $s - u \leq t \leq s + u$, or

(2) $s = a$ *and we can find a* $u > 0$ *such that* $f(t) \le f(a)$ *whenever* $a \le t \le a + u$, *or*

(3) $s = b$ *and we can find a* $u > 0$ *such that* $f(t) \le f(b)$ *whenever* $b - u \le t \le b$.

Exercise 2.5.6. *(i) Locate the local maxima for the curves of Figure 2.4.*

(ii) Explain why a global maximum must be a local maximum, but give an example of a local maximum which is not a global maximum.

(iii) Define what it means for f to have a local minimum and to have a global minimum at a point s.

(iv) Locate the local and global minima for the curves of Figure 2.4.

It frequently happens that $f(t)$ attains a global maximum at one of the *end points* a or b. An amplifier is loudest when the knob is turned full on.

Let us look at what happens at a point $a < c < b$ (that is to say, a point which is not an end point; we refer to such points as *interior points*). In this case, we know that, since we consider only the case when f is well behaved,

$$f(c + \delta t) = f(c) + f'(c)\delta t + o(\delta t).$$

If $f'(c) > 0$, then $f(c + \delta t) > f(c)$ when δt is strictly positive and sufficiently small, so f does not have a maximum at c. If $f'(c) < 0$, then $f(c + \delta t) > f(c)$ when δt is strictly negative and sufficiently small so f does not have a maximum at c. Thus, if f attains a local maximum at c, we must have $f'(c) = 0$.

Exercise 2.5.7. *Show that, if f attains a local minimum at an interior point c, we must have* $f'(c) = 0$.

Exercise 2.5.8. *Show that, if f attains a local maximum at the end point a, we must have* $f'(a) \le 0$.

State and prove a corresponding result when f attains a local maximum at the end point b.

Exercise 2.5.9. *Let* $a = -1$, $b = 1$ *and* $f(t) = t^3$. *Show that* $f'(0) = 0$, *but f does not attain a local maximum or minimum at 0.*

The result of Exercise 2.5.9 is disappointing, but we can still say something very useful. If c is an interior point with $f'(c) = 0$ and we can find a $u > 0$ such that $f'(t) \ge 0$ for $c - u \le t \le c$ and $f'(t) \le 0$ for $c \le t \le c - u$, then we know that $f(t)$ increases as t runs from $c - u$ to c and decreases as t runs from c to $c + u$. Thus f attains a local maximum at c.

Exercise 2.5.10. *State and prove a similar result about local minima at interior points.*

Exercise 2.5.11. *Show that, if we can find a $u > 0$ such that $f'(t) \leq 0$ for $a \leq t \leq a + u$, then f attains a local maximum at the end point a.*

State and prove the corresponding condition which ensures that f attains a local maximum at the end point b.

We thus have the following useful observation. Suppose that $k = 0$ or $k = 1$ and

(1) $a = x_0 < x_1 < x_2 \cdots < x_{n-2} < x_{n-1} < x_n < x_{n+1} = b$,
(2) $f'(x_j) = 0$ for $1 \leq j \leq n$,
(3) $(-1)^{j+k} f'(t) > 0$ for $x_j < t < x_{j+1}, 0 \leq j \leq n$.

Then f attains a local maximum at those x_j with $0 \leq j \leq n + 1$ and $j + k$ odd, f attains a local minimum at those x_j with $0 \leq j \leq n + 1$ and $j + k$ even and these are the only points where f attains a local maximum or minimum. The global maximum is attained at whichever point or points that f attains its largest local maximum.

Exercise 2.5.12. *Sketch an f obeying conditions (1), (2) and (3) with $n = 4$ and $k = 0$. Sketch an f obeying conditions (1), (2) and (3) with $n = 4$ and $k = 1$.*

Exercise 2.5.13. *Prove the statement made in the paragraph preceding Exercise 2.5.12.*

Exercise 2.5.14. *Suppose that conditions (1), (2) and (3) apply. Let $e_j = 1$ if f has a local maximum at x_j and $e_j = -1$ if f has a local minimum at x_j. Show that*

$$\tfrac{1}{2}e_0 + e_1 + e_2 + \cdots + e_{n-1} + e_n + \tfrac{1}{2}e_{n+1} = 0.$$

[This formula foreshadows the much more interesting result discussed in Section 8.3.]

Exercise 2.5.15. *Let $a = -1, b = 1$.*

(i) Show that, if $f(t) = 0$ for all t, f does not satisfy our conditions.
(ii) Show that, if $f(t) = t^3$ for all t, f does not satisfy our conditions.

Although the problems suggested by Exercise 2.5.15 cannot be dismissed, more likely difficulties are that we do not know how to differentiate f or, if we do, we have difficulties solving the equation $f'(t) = 0$ to find the x_j.

You should always remember that the mere fact that x satisfies $f'(x) = 0$ (the traditional name for such an x is a *stationary point*) does not tell you whether you have a local maximum or a local minimum (or, if you are

unlucky and stumble onto something resembling Exercise 2.5.15 (ii), neither[12]).
Thompson puts the matter in his usual trenchant manner

> ... this much-belauded process of equating to zero entirely fails to tell you whether
> the x that you thereby find is going to give you a maximum value of f or a
> minimum value of f. Quite so. It does not of itself discriminate; it finds for you the
> right value of x but leaves you to find out for yourselves whether the corresponding
> $f(x)$ is a maximum or a minimum. Of course, if you have plotted the curve, you
> know already which it will be.

To this I can only add the further warning that the global maximum may be at
an *end point* and not at a stationary point.

In the real world, it is important to think about the nature of the problem and
you should not be satisfied unless you understand what f looks like and why
the maximum occurs where it does.

2.6 Snell's law

Our next problem is of the same type as Problem A in the previous section, but
turns out to have a useful physical interpretation.

Problem B. *The beach at Camsea is the half plane consisting of those points
(x, y) with $y > 0$ and the ocean is the half plane consisting of those points
(x, y) with $y \leq 0$. Supergirl is sunbathing at the point A given by (a, b) when
she observes Superman being attacked by a shark at the point C given by (c, d).
(We take $a > c$, $b > 0 > d$.) Supergirl instantly decides to run directly to the
point X given by $(s, 0)$ and then swim directly to C. If Supergirl can run at
speed u and swim at speed v, what choice of s will get her to C as quickly as
possible?*

Exercise 2.6.1. *(i) Show that she takes a time*

$$f(s) = \frac{1}{u}\sqrt{(a - s)^2 + b^2} + \frac{1}{v}\sqrt{(c - s)^2 + d^2}.$$

(ii) Explain why $f(s)$ is large when $|s|$ is large.

Exercise 2.6.2. *Show that*

$$f'(s) = -\frac{1}{u}\frac{a - s}{\sqrt{(a - s)^2 + b^2}} - \frac{1}{v}\frac{c - s}{\sqrt{(c - s)^2 + d^2}}.$$

---

[12] This is more likely to happen in exams than in real life, for reasons given on page 126.

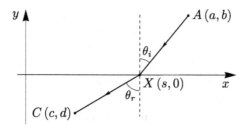

Figure 2.5 Snell's law

We have $f'(s) = 0$ when

$$\frac{1}{u} \frac{a - s}{\sqrt{(a - s)^2 + b^2}} = -\frac{1}{v} \frac{c - s}{\sqrt{(c - s)^2 + d^2}}$$

and simple trigonometry shows that this is equivalent to the statement that

$$\frac{1}{u} \sin \theta_i = \frac{1}{v} \sin \theta_r$$

or equivalently

$$\frac{\sin \theta_i}{\sin \theta_r} = \frac{u}{v}, \qquad \qquad \bigstar$$

where θ_i is the angle between the positive y axis and AX and θ_r is the angle between XB and the negative y axis. (Figure 2.5 may make this clearer. We need to be careful about the sign of the angle.)

Formula $\bigstar$ is called Snell's law. We can allow ourselves to bask in its elegance for a few minutes, but we are not quite done.

Observe that $\bigstar$ is not the only condition that θ_r and θ_i must satisfy. By simple trigonometry,

$$(a - s) = b \tan \theta_i, \quad (s - c) = d \tan \theta_r$$

and so

$$a - c = b \tan \theta_i + d \tan \theta_r.$$

We must check that this is compatible with Snell's law.

Rather than plunge into a thorny thicket of algebra, we proceed as follows. Let

$$g(s) = \frac{1}{u} \sin \theta_i - \frac{1}{v} \sin \theta_r.$$

We observe that $\sin \theta_i$ decreases strictly and $\sin \theta_r$ increases strictly as s increases from c to a. It follows that $g(s)$ increases strictly as s increases from c to a. When $s = c$, we have $\sin\theta_r = 0$ and $\sin\theta_i > 0$, so $g(c) > 0$, and, when $s = a$, we have $\sin\theta_i = 0$ and $\sin\theta_r > 0$, so $g(a) < 0$. Thus there exists a unique s_0 with $c < s_0 < a$ such that $g(s_0) = 0$.

Of course, the fact that $f'(s_0) = 0$ merely shows that s_0 is a stationary point. However, since s_0 is the unique stationary point, Exercise 2.6.1 (ii) shows that it must be the minimum.

If a ray of light crosses a plane boundary between two transparent media, then the angle of incidence θ_i and the angle of refraction θ_r obey Snell's law with u the speed of light in one medium and v the speed of light in the other. We thus have an example of Fermat's principle that 'light follows the fastest path.' Here is another example.

Exercise 2.6.3. *Supergirl is milking a cow at the point A with coordinates (a, b). Suddenly she observes Superman being attacked by a small fire-breathing dragon at the point C with coordinates (p, q). (We take b, $q > 0$.) She instantly seizes a large pail and runs in a straight line from A to a brook flowing along the x-axis, reaching it at a point X given by $(x, 0)$, fills the pail with water and then runs in a straight line to the point C where she throws the water over the dragon.*

Show that, if she runs at constant speed and wishes to take the least possible time, she will choose X so that AX and XC make equal angles with the y-axis. In the language of light rays, the angle of incidence equals the angle of reflection. (See page 25.)

[No dragons were harmed in the making of this exercise.]

Exercise 2.6.4. *A rectangle has perimeter $4a$. If one side has length x, find the area $f(x)$ in terms of x. Show that, among all rectangles of given perimeter, the squares have greatest area.*

Exercise 2.6.5. *Recall Heron's formula from Example 1.2.6. Show that, among the triangles with fixed perimeter $2s$ and one side of fixed length c, the triangles with the other two sides of equal length have greatest area.*

Show that, among the triangles with fixed perimeter, the equilateral triangles (that is to say, the triangles with all sides of equal length) have greatest area.

Exercise 2.6.6. *(i) Let n be an integer with $n \geq 2$ and let $a > 0$. If*

$$f(x) = x\left(\frac{na - x}{n - 1}\right)^{n-1},$$

show that

$$f(a) > f(x)$$

for all x with $na \geq x \geq 0$ and $x \neq a$.

(ii) Observe that when $n = 2$ we recover the result of Exercise 2.6.4.

(iii) Show, using (i) when $n = 2$ and 3, that, if x, y, $z \geq 0$ and $x + y + z = 3a$, then $a^3 \geq xyz$ with equality if and only if $x = y = z$.

(iv) Generalise the result of (iii) to n variables and prove it.

(v) Obtain Cauchy's arithmetic-geometric inequality

$$\frac{x_1 + x_2 + \cdots + x_n}{n} \geq (x_1 x_2 \ldots x_n)^{1/n}$$

for $x_j \geq 0$.

Under what circumstances do we have

$$\frac{x_1 + x_2 + \cdots + x_n}{n} = (x_1 x_2 \cdots x_n)^{1/n}?$$

3

Functions, old and new

3.1 The logarithm

If we wish to find the value of

$$G(x) = \int_0^x \frac{t}{(1+t^2)^2}\, dt,$$

then a little thought suggests that we look at

$$F(t) = \frac{-1}{2(1+t^2)}.$$

Exercise 3.1.1. *Use our rules for differentiation to show that*

$$F'(t) = \frac{t}{(1+t^2)^2}$$

and deduce, using the fundamental theorem of the calculus, that $G(x) = F(x)$.

However, just as the calculus introduces new ideas, so it produces a host of new functions. Consider the function

$$l(x) = \int_1^x \frac{1}{t}\, dt$$

for $x > 0$. The method of the paragraph above suggests trying to find polynomials P and Q with no common factor such that, writing $L(t) = P(t)/Q(t)$, we have $L'(t) = 1/t$. Using the standard rules for differentiation to find $L'(t)$, we get

$$\frac{P'(t)}{Q(t)} - \frac{P(t)Q'(t)}{Q(t)^2} = \frac{1}{t}$$

and so, multiplying out,

$$t\big(P'(t)Q(t) - Q'(t)P(t)\big) = Q(t)^2. \qquad\qquad ★$$

Exercise 3.1.2. *Check these statements.*

Since ★ is an equation only involving polynomials which holds for all $t > 0$ it must hold for all t. Setting $t = 0$, we see that $Q(0) = 0$ and so t must be a factor of $Q(t)$. Thus $Q(t) = t^m R(t)$, where $m \geq 1$ and t is not a factor of $R(t)$. Substituting in equation ★, we get

$$t\left(t^m P'(t)R(t) - t^m R'(t)P(t) - mt^{m-1} R(t)P(t)\right) = t^{2m} R(t)^2$$

and so

$$t P'(t)R(t) - t R'(t)P(t) - mR(t)P(t) = t^m R(t)^2.$$

Setting $t = 0$, once more, we get $R(0)P(0) = 0$, so $P(0) = 0$ and t must be a factor of $P(t)$. Thus t is a factor both of $P(t)$ and of $Q(t)$, contradicting our initial assumption that $P(t)$ and $Q(t)$ have no common factor. We have shown that l is not the ratio of two polynomials. (The reader may wish to compare the standard demonstration that $\sqrt{2}$ is irrational given in Exercise 7.3.1.)

The fact that l is a new type of function does not mean that we cannot discover many of its properties. Thus, for example, the fundamental theorem of the calculus tells us that

$$l'(x) = \frac{1}{x}.$$

Exercise 3.1.3. *Explain the meaning of $l(a)$ in terms of the area under a certain curve for $a \geq 1$ and for $1 > a > 0$.*

Our next result is central. If u, $v > 0$, then (by Exercise 2.2.10 (iii))

$$l(uv) = \int_1^{uv} \frac{1}{t}\, dx = \int_1^u \frac{1}{t}\, dt + \int_u^{uv} \frac{1}{t}\, dt.$$

Now the change of variable formula given on page 45 (with $a = 1$, $b = u$, $f(t) = 1/t$ and $G(t) = ut$) yields

$$\int_u^{uv} \frac{1}{t}\, dt = \int_1^v u \times \frac{1}{ut}\, dt = \int_1^v \frac{1}{t}\, dt = l(v)$$

and so

$$l(uv) = l(u) + l(v).$$

We see that the function l converts multiplication into addition. It is a variation on the 'logarithm' invented by Lord Napier and improved by Briggs. Before the age of electronic calculators, it was much easier to do addition than multiplication and, for 300 years, the logarithm was an essential calculating tool.

There is a famous story in William Lilly's *History of his Life and Times* about the first meeting of these two benefactors of mankind.

> Mr Briggs appoints a certain day when to meet at Edinburgh; but failing thereof, [Lord Napier] was fearful he would not come. It happened one day as John Marr and the Lord Napier were speaking of Mr Briggs, 'Oh! John,' saith [Lord Napier], 'Mr Briggs will not come now'; at the very instant one knocks at the gate, John Marr hastened down and it proved to be Mr Briggs to his great contentment. He brings Mr Briggs into my Lord's chamber, where almost one quarter of an hour was spent, each beholding the other with admiration, before one word was spoken. At last Mr Briggs began, – 'My Lord, I have undertaken this long journey purposely to see your person, and to know by what engine of wit or ingenuity you came first to think of this most excellent help unto astronomy, viz. the Logarithms; but, my Lord, being by you found out, I wonder nobody else found it out before, when now known it is so easy.'

Although the author of the present book is old enough to have been thoroughly trained in the *computational* use of the logarithm, the method is now obsolete. However, the logarithm retains its *mathematical* importance. From now on, we shall write $\log x = l(x)$. The reader should be prepared to meet the notations $\log_e x = \ln x = l(x)$, but these are hangovers from the computational past of the logarithm.[1]

Exercise 3.1.4. *(i) Use the relation* $\log(ab) = \log a + \log b$ *(for* a, $b > 0$*) to show that* $\log 1 = 0$ *and* $\log a = -\log(1/a)$.

(ii) Use the definition $\log x = \int_1^x t^{-1}\, dt$ *to show that* $\log 1 = 0$. *Use the definition and the change of variable formula to show that* $\log a = -\log(1/a)$.

Exercise 3.1.5. *(i) Explain why* $\log x$ *is a strictly increasing function of* x.

(ii) Use the fact that $2^{-1} \leq t^{-1}$ *for* $1 \leq t \leq 2$ *to show that* $(x-1)/2 \leq \log x$ *for* $1 \leq x \leq 2$.

(iii) Show that $\log 2^n = n \log 2$ *and deduce that, given any* K, *we can find an* n *such that* $\log 2^n > K$. *(Thus* $\log x$ *increases without bound as* x *increases.)*

(iv) Show that $\log x$ *decreases without bound as* x *decreases towards* 0.

It is interesting to note that, although $\log x$ increases without bound, it does so very slowly. Suppose that n is a strictly positive integer. Then, if $f(x) = x^{-1/n} \log x$, the standard rules of differentiation yield

$$f'(x) = -\frac{1}{n}x^{-1-1/n}\log x + x^{-1/n}x^{-1} = \frac{1}{n}x^{-1-1/n}(n - \log x).$$

---

[1] I have not emphasised the point, but the reader should always remember that $\log x$ is only defined if $x > 0$.

We know that there is a K_n such that $\log x > n$ whenever $x > K_n$, so $f'(x) < 0$ and $f(x)$ is decreasing for $x > K_n$. Thus, $\log x$ increases slower than $x^{1/n}$ when x is large. This gives another demonstration that $\log x$ is not the ratio of two polynomials.

Exercise 3.1.6. *Let us set* $f(x) = \log(\log x)$ *when x is sufficiently large for the expression to make sense. Explain why $f(x)$ increases without bound as x increases. Guess a value of K such that $\log \log K > 4$ and check if your answer is correct by using a calculator.*[2]
[We continue this chain of thought in Exercise 3.2.6.]

Exercise 3.1.7. *In* Calculus Made Easy, *Thompson writes*[3] *'The solution of a differential equation often seems as different from the original expression as a butterfly does from the caterpillar that it was. Who would have supposed that such an innocent thing as*

$$f'(x) = \frac{1}{a^2 - x^2}$$

(for $|x| < a$) could blossom out into

$$f(x) = \frac{1}{2a} \log \frac{a + x}{a - x} + C$$

(where C is a constant) yet the latter is the solution of the former.'

Verify that, if f is given by the second formula, then f' is given by the first one.

3.2 The exponential function

We now look at the equation

$$\log x = y. \qquad \bigstar$$

Since $\log t$ is a strictly increasing function of t, we know that equation $\bigstar$ has at most one solution. On the other hand, we know that $\log t$ increases unboundedly as t increases and decreases unboundedly as t decreases towards 0. Thus we can find positive numbers K_1 and K_2 such that

$$\log K_1 < y - 1 \text{ and } y + 1 < \log K_2.$$

It follows that we can find an x lying between K_1 and K_2 such that $\log x = y$.

---

[2] The logarithm function on your calculator is probably labelled ln or $\log_e$. (The button labelled log or $\log_{10}$ gives a different, though related, function.)
[3] I have slightly modified the quotation to be consistent with this text.

We now know that equation ★ has a unique solution which we shall call $E(y)$. By definition, $\log E(y) = y$ whilst

$$\log \big(E(\log x) \big) = \log x$$

so $E(\log x) = x$. Looking at Definition 1.4.6, we see that E is the inverse function of log. It seems clear that $\log x$ and $E(x)$ are sufficiently well behaved functions of x to allow us to use the inverse function rule and obtain the remarkable formula

$$E'(x) = \frac{1}{\log'(E(x))} = \frac{1}{\frac{1}{E(x)}} = E(x).$$

It is easy to derive many properties of E from the fact that it is the inverse function of log.

Exercise 3.2.1. *(i) Show that, if a, $b > 0$, then*

$$E\big(\log a + \log b \big) = E\big(\log a \big) \times E\big(\log b \big)$$

and deduce that $E(x + y) = E(x)E(y)$ for all x and y.
(ii) Show that $E(0) = 1$ and $1/E(x) = E(-x)$.
(iii) Explain why $E(x) > 0$ for all x.
(iv) Show that $E(x)$ is a strictly increasing function of x.
(v) Show that, given $a > 0$, we can find a K such that $E(x) > a$ whenever $x > K$. Show also that, given $b > 0$, we can find an L such that $b > E(x) > 0$ whenever $L > x$.

We call $E(x)$ the *exponential function* and write

$$\exp x = E(x).$$

As might be expected from the fact that $\log x$ increases very slowly, $\exp x$ is a very rapidly increasing function of x when x is large.

Exercise 3.2.2. *Let n be a positive integer. Show that $x^{-n} \exp x$ is an increasing function of x for $x \geq n$.*

Exercise 3.2.3. *(This exercise is taken from* Calculus Made Easy. *Kelvin designed the first successful trans-Atlantic cable. The value of s represented the difference between financial triumph and disaster.)*
    It was shown by Lord Kelvin that the speed of signalling through [an undersea] cable depends on the value of the ratio of the external diameter of the core to the diameter of the enclosed copper wire. If this ratio is called y, then the number of signals s that can be sent per minute can be expressed by

the formula

$$s = ay^2 \log \frac{1}{y},$$

where a is a quantity depending on the length and quality of the materials. Show that s will be a maximum if $y = \exp(-1/2)$.

Once we have log and exp, we can construct other interesting functions. We start with a preliminary exercise.

Exercise 3.2.4. *Let* $x > 0$.

(i) *Explain why* $n \log x = \log x^n$ *for n a positive integer.*
(ii) *Deduce, or show otherwise, that* $n \log x = \log x^n$ *for all integers n.*
(iii) *Deduce that*

$$\frac{p}{q} \log x = \log x^{p/q}$$

and so

$$\exp\left(\frac{p}{q} \log x\right) = x^{p/q}$$

for all integers p, q with $q \neq 0$.

Thus, if we *define*

$$x^a = \exp(a \log x)$$

whenever $x > 0$, our new definition will be consistent with our old one when a is rational (that is to say, when $a = p/q$ for integers p, q with $q \neq 0$). It is easy to check that our extended definition works as we might hope.

Exercise 3.2.5. *Let* $x, y > 0$ *and let a and b be constants. Prove the following results.*

(i) $x^{a+b} = x^a x^b$.
(ii) $(xy)^a = x^a y^a$.
(iii) $x^{ab} = (x^a)^b$.
(iv) *If we write* $e = \exp 1$, *then* $\exp a = e^a$.
(v) *Let* $f(t) = t^a$ *for* $t > 0$. *Then* $f'(t) = at^{a-1}$.

From now on we follow general usage and sometimes write $\exp a$ and sometimes e^a.

Exercise 3.2.6. *Explain why there is an M such that*

$$\log \log \log M \geq 4.$$

Could you write down all the digits of such an M in a lifetime?
[*Hint:* $e^5 > 100$.]

We also get a new result which, like Exercise 3.2.5 (v), is a consequence of the function of a function rule.

Exercise 3.2.7. *Suppose that a is a constant with $a > 0$. Show that, if we set $g(t) = a^t$, then*

$$g'(t) = (\log a)a^t.$$

Exercise 3.2.8. • *Let $f(t) = t^{1/t}$ for $t > 0$. Find $f'(t)$, show that f has a unique maximum, find that maximum and sketch the graph of f. Show that the equation*

$$n^m = m^n$$

with $n > m$ has exactly one solution with n and m strictly positive integers. [*Hint: In Section 6.3 we shall show that $e \approx 2.7$.*]

Many students and some of their teachers object to our way of defining x^a. Surely, they say, it would be more natural to define x^a by saying that we know what $x^{p/q}$ ought to be and we choose the values of x^a when a is not rational (i.e. not of the form p/q) to 'fill the gaps in a natural manner'.

If you agree,[4] you should, at least, consider how we might calculate x^a.
Using our definition. Compute $\log x$, $a \log x$ and $\exp(a \log x)$.
Filling in the gaps. Choose integers p and q with $q \geq 1$ so that p/q is 'sufficiently close' to a. Now compute the pth power a^p and the qth root $a^{p/q}$ (note that you cannot use the x^a button on your calculator, since your calculator uses the $\exp(a \log x)$ definition). Provided we chose p/q correctly, $a^{p/q}$ will be 'sufficiently close' to a^x.

Exercise 3.2.9. *Think about the two proposed methods when $x = a = \pi$.*

If the reader has done a bit of probability, she may be aware that

$$n! = 1 \times 2 \times 3 \times \cdots \times n$$

(pronounced 'n factorial') plays an important role. (See, for example, Section 5.3.) As the subject develops, it becomes necessary to have good estimates for $n!$ when n is very large. The ideas that follow are due to de Moivre,[5] but were developed by Stirling and Euler.

---

[4] If your teachers agree, they should undertake the amusing and informative task of proving Exercise 3.2.5 (v) by 'filling the gaps'. They should also read the discussion of logarithms in Klein's magnificent *Elementary Mathematics from an Advanced Standpoint* [4].
[5] In later life, Newton used to put off questions on the *Principia Mathematica* with the words 'Go to Mr de Moivre; he knows these things better than I do.'

The first step is to look at $\log n!$ rather than $n!$. We then have

$$\log n! = \log 1 + \log 2 + \cdots + \log n.$$

We have already used the next idea in Exercise 2.2.11.

Observe that

$$\log r \le \log t \le \log(r + 1)$$

for $r \le t \le r + 1$ and so

$$\log r = \int_r^{r+1} \log r \, dt \le \int_r^{r+1} \log t \, dt \le \int_r^{r+1} \log(r + 1) \, dt = \log(r + 1).$$

Adding up these inequalities, we get

$$\log 1 + \log 2 + \cdots + \log n \le \int_1^2 \log t \, dt + \int_2^3 \log t \, dt + \cdots + \int_n^{n+1} \log t \, dt$$

$$\le \log 2 + \log 3 + \cdots + \log(n + 1)$$

and so, simplifying,

$$\log n! \le \int_1^{n+1} \log t \, dt \le \log(n + 1)!.$$

The last expression is not quite in the form we want, but it is easy to check that it implies

$$\int_1^n \log t \, dt \le \log n! \le \int_1^{n+1} \log t \, dt.$$

To evaluate the integrals we use a standard (but rather clever) application of integration by parts.

Exercise 3.2.10. *By setting $F(t) = t$ and $G(t) = \log t$ in the formula for integration by parts (see page 44), show that*

$$\int_1^n \log t \, dt = n \log n - (n - 1).$$

We now tidy up.

Exercise 3.2.11. *(i) Explain why*

$$\int_1^{n+1} \log t \, dt \le \int_1^n \log t \, dt + \log(n + 1).$$

(ii) Use the results so far obtained to show that

$$(n \log n) - (n - 1) \le \log n! \le (n \log n) - (n - 1) + \log(n + 1).$$

(iii) Use a calculator to find what the inequality of (ii) becomes when we set $n = 1000$.

(iv) (A bit of a by-way.) $10! = 362\,8800$ *ends in two zeros. How many zeros are there at the end of* $1000!$?

(v) By observing that $10^m \le r < 10^{m+1}$ *implies*

$$m \le \frac{\log r}{\log 10} < m + 1$$

and vice versa, estimate the number of digits in $1000!$ *when written out in full.*

(vi) By applying the function exp *to the inequality in (ii), show that*

$$n^n e^{1-n} \le n! \le (n+1)n^n e^{1-n},$$

where $e = \exp 1$.

It is not hard to refine these ideas. The next exercise is quite long and will not be needed later.

Exercise 3.2.12.• *(i) In our calculations, we compared* $\log r$ *with* $\int_{r-1}^{r} \log t \, dt$ *and* $\int_{r}^{r+1} \log t \, dt$. *By drawing a diagram, or otherwise, convince yourself that it might give better results to compare* $\log r$ *with* $\int_{r-1/2}^{r+1/2} \log t \, dt$.

(ii) Let

$$f(t) = 2\log r - \big(\log(r - t) + \log(r + t)\big).$$

By looking at $f'(t)$, *show that*

$$0 \le f(t) \le \frac{1}{2}\left(\frac{1}{r - \frac{1}{2}} - \frac{1}{r + \frac{1}{2}}\right)$$

for $0 \le t \le 1/2$.

(iii) Show that

$$\int_0^{1/2} f(t)\,dt = \log r - \int_{r-1/2}^{r+1/2} \log t \, dt$$

and deduce that

$$0 \le \log r - \int_{r-1/2}^{r+1/2} \log t \, dt \le \frac{1}{2}\left(\frac{1}{2r - 1} - \frac{1}{2r + 1}\right).$$

(iv) By adding these inequalities, show that

$$0 \le \log n! - \int_{1/2}^{n+1/2} \log t \, dt \le \frac{1}{2}.$$

Deduce that

$$(n + \tfrac{1}{2})\log(n + \tfrac{1}{2}) - \tfrac{1}{2}\log\tfrac{1}{2} - n$$
$$\leq \log n! \leq (n + \tfrac{1}{2})\log(n + \tfrac{1}{2}) - \tfrac{1}{2}\log\tfrac{1}{2} - n + \tfrac{1}{2}$$

and

$$2^{-1/2}e^{-n}(n + \tfrac{1}{2})^{(n+\frac{1}{2})} \leq n! \leq e^{1/2}2^{-1/2}e^{-n}(n + \tfrac{1}{2})^{(n+\frac{1}{2})}.$$

(v) *If you are not already exhausted, find the appropriate inequalities if we use the approximation*

$$\log n! \approx \log m! + \int_{m+1/2}^{n+1/2} \log t \, dt$$

in place of $\log n! \approx \int_{1/2}^{n+1/2} \log t \, dt$.

Exercise 3.2.13.• *Use the methods of this section to obtain the following results.*

(i) *If* $0 \geq a > -1$, *then*

$$\left|(1^a + 2^a + \cdots + n^a) - \frac{n^{a+1}}{a + 1}\right| \leq \frac{1}{a + 1}.$$

(ii) $|(1^{-1} + 2^{-1} + \cdots + n^{-1}) - \log n| \leq 1.$

(iii) *If* $a \geq 0$, *then*

$$\left|(1^a + 2^a + \cdots + n^a) - \frac{n^{a+1}}{a + 1}\right| \leq n^a.$$

(*We have already proved this when a is rational by a similar method in Exercise 2.3.5.*)

(iv) *What happens if we try and apply our method when* $-1 > a$? *Why is the result useless?* (*Actually, if we thought more deeply, we could extract useful information, but we shall not do this.*)

3.3 Trigonometric functions

The exponential function will play a major role in several of the chapters that follow. In contrast, we shall use the trigonometric functions only as a source of further examples. The reader can thus pay less attention to this section than to the two previous sections.

If I compute sin 30 on my scientific calculator, I will get different answers according to whether I have set my calculator to degrees, grads or radians. The degree is a unit of angle which can be traced back to the Ancient Babylonian

Functions, old and new

astronomers. There are 90 degrees in one right angle. The grad is a relic of an attempt to decimalise angular measurement and is sometimes used in surveying. There are 100 grads in one right angle. Those who take pleasure in such things record the brad linked to the binary system and the mil (see the footnote on page 68). There are 64 brads and 1600 mils in one right angle.

Because different people use different systems, we shall write $\sin_A$ and $\cos_A$ for the value of the sine and cosine functions when angles are measured in a system A in which there are k_A units in a right angle. We shall assume that $\sin_A$ and $\cos_A$ are well behaved and seek to find the derivative of $\sin_A$.

Exercise 3.3.1. *Explain why*

$$\sin_A(k_A t) = \sin_B(k_B t)$$

and deduce that $k_A \sin'_A(k_A t) = k_B \sin'_B(k_B t)$. Conclude that

$$k_A \sin'_A 0 = k_B \sin'_B 0.$$

We shall use the $o(\delta t)$ notation from Section 1.4. The addition formula for the sine function yields

$$\sin_A(t + \delta t) = \sin_A t \cos_A \delta t + \cos_A t \sin_A \delta t.$$

The definition of the derivative yields

$$\cos_A \delta t = \cos_A 0 + (\cos'_A 0) \times \delta t + o(\delta t) = 1 + (\cos'_A 0) \times \delta t + o(\delta t)$$

and

$$\sin_A \delta t = \sin_A 0 + (\sin'_A 0) \times \delta t + o(\delta t) = (\sin'_A 0) \times \delta t + o(\delta t).$$

Since $\cos_A t$ has a maximum when $t = 0$, we know that $\cos'_A 0 = 0$ and so

$$\cos_A \delta t = 1 + o(\delta t).$$

Putting these results together, we get

$$\sin_A(t + \delta t) = \sin_A(t) + (\cos_A t) \times (\sin'_A 0)\delta t + o(\delta t)$$

and so $\sin'_A t = (\sin'_A 0) \times (\cos_A t)$.

Exercise 3.3.1 tells us that we can find a system R for which $\sin'_R 0 = 1$ and our formula takes the particularly simple form

$$\sin'_R t = \cos_R t.$$

We call the unit of angle in this system the *radian*. However, we still have to find k_R.

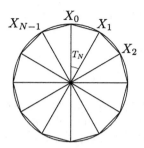

Figure 3.1 Cake cutting again

To this end, we look again at our discussion of π which started on page 30. We consider a disc with centre O and radius 1. Divide the circumference into N equal parts using points $X_0, X_1 \ldots, X_{N-1}$ as shown in Figure 3.1.

Exercise 3.3.2. *(i) Use elementary trigonometry and a doubling formula from Exercise 1.5.3 (iv) to show that the area of the triangle with vertices O, X_0 and X_1 is*

$$\sin_R \frac{T_N}{2} \cos_R \frac{T_N}{2} = \frac{1}{2} \sin_R T_N,$$

where T_N is the angle (measured in radians) between OX_0 and OX_1.

(ii) Explain why $T_N = 4k_R/N$ and so the polygon $X_0, X_1 \ldots X_{N-1}$ has area

$$P_N = \frac{N}{2} \sin_R \frac{4k_R}{N}.$$

(iii) Explain why $\sin_R \delta t = \delta t + o(\delta t)$ and deduce that P_N gets arbitrarily close to $2k_R$ as N increases.

(iv) Conclude that $2k_R = \pi$.

Exercise 3.3.3. *(i) In the previous exercise we used the fact that the area of the polygon $X_0, X_1 \ldots X_{N-1}$ approaches the area enclosed by the escribed circle as N increases to show that $k_R = \pi/2$. Produce a similar demonstration of the same result using the fact that the length of the perimeter of the polygon approaches the length of the perimeter of the escribed circle.*

(ii) Suppose that A and B are points on the circumference of a circle of radius a with centre 0. If θ is the angle between OA and OB measured in radians (and $0 \le \theta < 2\pi$), show that the length of the arc subtended by the angle (that is to say, the appropriate arc between A and B) is $a\theta$.

Exercise 3.3.4.• **[Dipping distance for lighthouses]** *(This is an exercise in first order approximation.)*

(i) *Show that the distance that the top of a lighthouse, at height h above sea level, can be seen from the crow's nest of a ship, at height H above sea level, is approximately*

$$K(\sqrt{h} + \sqrt{H})$$

where K is a constant depending on the radius of the Earth.

(ii) *A semaphore company wishes to establish a chain of towers across a wide desert in such a way that the top of each tower is visible from the tops of its neighbours. Discuss whether it is better to build a few very tall towers or many shorter ones.*

The fact that $\pi/2$ is irrational makes radians unsuitable for many practical applications. However, the simplicity of the relation $\sin'_R t = \cos_R t$ means that radians are the only angular measure used in any situation involving calculus. We shall write $\sin t = \sin_R t$ and $\cos t = \cos_R t$ from now on.[6]

Exercise 3.3.5. *(i) Use the fact that $\cos t = \sin(t + \frac{\pi}{2})$ and $\sin t = -\cos(t + \frac{\pi}{2})$ to show that $\cos' t = -\sin t$.*

(ii) Use the standard rules for differentiation to show that (for appropriate values of t)

$$\tan' t = (\sec t)^2, \ \cot' t = -(\operatorname{cosec} t)^2,$$

$$\operatorname{cosec}' t = -\operatorname{cosec} t \cot t \ \text{and} \ \sec' t = \sec t \tan t.$$

Exercise 3.3.6. *Supergirl is at $(a, 0)$ reading an improving book by the side of circular pool of radius a with centre at $(0, 0)$ when she observes Superman being attacked by a giant wasp at $(-a, 0)$. Supergirl instantly decides to run round the edge of the pool to the point $(a \cos \theta, a \sin \theta)$, where θ is measured in radians, and then to dive in and swim directly to $(-a, 0)$. If Supergirl can run at speed u and swim at speed v, show that this will take her a time*

$$\frac{a\theta}{u} + \frac{2a}{v} \cos \frac{\theta}{2}.$$

Show that Supergirl will either dive in at once (that is to say, take $\theta = 0$) or run all the way round (that is to say, take $\theta = \pi$).

Exercise 3.3.7.• *(This is just an exercise in differentiation and will not be used later.)*

---

[6] The approximation $\sin t \approx t$ (for t small) is so useful in gunnery that the military have adopted an angular measure which is almost (but not quite) one thousandth of a radian. There are 1600 *mils* in a right angle and $1600 \approx 10^3 \times (\pi/2)$.

(i) *Show that* $\sin x$ *is strictly increasing as* x *runs from* $-\pi/2$ *to* $\pi/2$. *Thus it appears (and is, indeed, true) that* $\sin$ *has an inverse function* $\sin^{-1} y$ *defined for* $-1 \le y \le 1$. *Use the trigonometric formula* $(\sin t)^2 + (\cos t)^2 = 1$ *to obtain*

$$\cos(\sin^{-1} y) = \sqrt{1 - y^2}.$$

Use the inverse function rule to obtain

$$(\sin^{-1})'(y) = \frac{1}{\sqrt{1 - y^2}}.$$

(ii) *Show that* $\cos x$ *is strictly decreasing as* x *runs from* 0 *to* π. *Show that, if we define the appropriate inverse function,*

$$(\cos^{-1})'(y) = \frac{-1}{\sqrt{1 - y^2}}.$$

(iii) *Show that* $\tan x$ *is strictly increasing as* x *runs from* $-\pi/2$ *to* $\pi/2$ *(but excluding* $x = -\pi/2$ *and* $\pi/2$*). Show that, if we define the appropriate inverse function,*

$$(\tan^{-1})'(y) = \frac{1}{1 + y^2}.$$

Exercise 3.3.8.• *(Another exercise that will not be used later.) In this exercise we study the properties of the functions*[7]

$$\cosh x = \frac{\exp(x) + \exp(-x)}{2} \text{ and } \sinh x = \frac{\exp(x) - \exp(-x)}{2}.$$

(i) *Show that* $\cosh(-x) = \cosh x$, $\sinh(-x) = -\sinh x$.
(ii) *Show that* $\cosh' x = \sinh x$, $\sinh' x = \cosh x$.
(iii) *Show that* $\sinh x$ *is strictly increasing as* x *increases. Show that* $\cosh x$ *is strictly increasing as* x *increases from* 0.
(iv) *Sketch the graphs of* $\cosh x$ *and* $\sinh x$ *separately.*
(v) *Sketch* $\cosh x$, $\sinh x$ *and* $(\exp x)/2$ *on the same graph.*
(vi) *Show that*

$$\cosh(x + y) = \cosh x \cosh y + \sinh x \sinh y,$$

$$\sinh(x + y) = \sinh x \cosh y + \cosh x \sinh y$$

---

[7] We pronounce cosh as written and sinh as 'sinch' or 'shine' according to local custom. The word cosh is an abbreviation for 'hyperbolic cosine' and the word sinh for 'hyperbolic sine'. The strong (but *not* perfect) resemblance between the various formulae for 'hyperbolic' and 'trigonometric' functions is not accidental.

and

$$\cosh^2 x - \sinh^2 x = 1$$

for all x and y.

Exercise 3.3.9.• The following exercise is close to the author's heart.

(i) Show that

$$\int_a^b \cos mt \, dt = \frac{\sin mb - \sin ma}{m}$$

if $m \neq 0$.

(ii) Use an addition formula (see Exercise 1.5.3) to show that

$$2 \cos u \cos v = \cos(u + v) + \cos(u - v).$$

(iii) Hence, or otherwise, show that, if n and m are integers, then

$$\int_{-\pi}^{\pi} \cos nx \cos mx \, dx = \begin{cases} 0 & \text{if } n \neq m, \\ \pi & \text{if } n = m \neq 0, \\ 2\pi & \text{if } n = m = 0. \end{cases}$$

4

Falling bodies

4.1 Galileo

At the age of 69, Galileo was forced into a humiliating public denial of the Copernican ideas that he had done so much to defend. 'Vehemently suspect of heresy', he was placed under house arrest for the rest of his life. Fearing that his ideas and discoveries would die with him, Galileo decided to write an account of his earlier researches.

The resulting book, published in Holland outside the reach of the Roman Inquisition, was entitled *Discourses and Mathematical Demonstrations Relating to Two New Sciences*[1] and was one of the most influential books ever written. The first new science was what we would now call material science and the second was what we now call dynamics.

> My purpose is to set forth a very new science dealing with a very ancient subject. There is, in nature, perhaps nothing older than motion, concerning which the books written by philosophers are neither few nor small; nevertheless I have discovered by experiment some properties of it which are worth knowing and which have not hitherto been either observed or demonstrated. Some superficial observations have been made, as, for instance, that the free motion of a heavy falling body is continuously accelerated; but to just what extent this acceleration occurs has not yet been announced; for so far as I know, no one has yet pointed out that the distances traversed, during equal intervals of time, by a body falling from rest, stand to one another in the same ratio as the odd numbers beginning with unity.
>
> It has been observed that missiles and projectiles describe a curved path of some sort; however, no one has pointed out the fact that this path is a parabola. But this and other facts, not few in number or less worth knowing, I have succeeded in proving; and what I consider more important, there have been opened up to this

---

[1] Galileo would have preferred the snappier title *Dialogues on Motion*, but, after the manner of their kind, his publishers chose this one.

vast and most excellent science, of which my work is merely the beginning, ways and means by which other minds more acute than mine will explore its remote corners.

Discourses Relating to Two New Sciences

Galileo was an able mathematician and a great theoretical physicist,[2] but he was, above all, a great observer and experimentalist.

> What you refer to is the method [Aristotle] uses in writing his doctrine, but I do not believe it to be that with which he investigated it. Rather I think it certain that he first obtained it by means of the senses, experiments and observations, to assure himself as much as possible of his conclusions. . . . And you may be sure that Pythagoras, long before he discovered the proof [of his theorem], was sure that the square on the side opposite the right angle in a right triangle was equal to the squares on the other two sides. . . . Aristotle . . . said many times [that he] preferred sensible experience to any argument.

Dialogue Concerning the Two Chief World Systems

At one time, it was fashionable to deny that Galileo did the experiments which he describes with such care and pride. Stillman Drake brought the lengthy controversy to an end by consulting Galileo's manuscripts and observing that they contained work sheets on which Galileo recorded some of his experimental data. It seems reasonable to suppose that Galileo used experiment to suggest theory and theory to suggest experiment and that his *Discourses* record his successful theories and experiments omitting those which led nowhere.

This is a book on mathematics and not on physics or the history of science. I shall simply write down Galileo's laws of motion in the modern notation of the calculus and show, using the methods of this book, how they lead to the conclusions he describes.

Galileo first considers a body falling vertically so that it is at a height $y(t)$ at time t. He states that

$$y''(t) = -g, \qquad \bigstar$$

where g is a positive constant.[3] Our problem is to discover $y(t)$. (In the jargon of the trade, we wish to *solve* the *differential equation* $\bigstar$.)[4] We do this in two stages by first finding the velocity $v(t) = y'(t)$ and then using this result to find $y(t)$.

---

[2] He also wrote better Italian, had a better view of the dangers of theological and philosophical over-reach and made better jokes than his opponents.

[3] We call $y'(t)$ the *velocity* and $y''(t)$ the *acceleration*. The equation $\bigstar$ thus tells us that a falling body has constant acceleration. Human beings have a deeply ingrained belief that by naming something we acquire power over it.

[4] Although I shall not supply any definition of these terms, it worth noting that the solution of a differential equation is not a number but a *function*.

Since $y''(t) = -g$, it follows that

$$v'(t) = -g$$

and a little thought suggests looking at the new function $f(t) = v(t) + gt$. The standard rules for differentiation yield

$$f'(t) = v'(t) + g = 0,$$

so f is constant, that is to say, we have $f(t) = A$ and

$$v(t) = A - gt$$

for some constant A. At first sight, the intrusion of a constant seems unwelcome, but, setting $t = 0$, we see that $v(0) = A$, so that A is the velocity at time $t = 0$. Thus $v(t) = v(0) - gt$ and we have to solve

$$y'(t) = -gt + v(0).$$

This time we have to think a little longer, but after some experimentation,[5] we look at $F(t) = y(t) + \frac{gt^2}{2} - v(0)t$ and observe that

$$F'(t) = y'(t) + gt - v(0) = 0,$$

so we have $F(t) = B$ and

$$y(t) = B - \frac{gt^2}{2} + v(0)t.$$

Setting $t = 0$, we see that $y(0) = B$ and so

$$y(t) = y(0) + v(0)t - \frac{gt^2}{2}$$

or, equivalently,

$$y(t) = y(0) + y'(0)t - \frac{gt^2}{2}.$$

Exercise 4.1.1. *Suppose that $y(0) = y'(0) = 0$. If $T > 0$ and M and N are positive integers, show that*

$$\frac{y((N+1)T) - y(NT)}{y((M+1)T) - y(MT)} = \frac{2N+1}{2M+1},$$

that is to say 'the distances traversed, during equal intervals of time, by a body falling from rest, stand to one another in the same ratio as the odd numbers beginning with unity.'

---

[5] Once again we *guess* the answer and then *check* that it is correct. Only routine problems can be solved in a routine way.

Exercise 4.1.2. Let $P(t) = a_0 + a_1 t + \cdots + a_n t^n$.

(i) Show that $y''(t) = P(t)$ has the general solution

$$y(t) = c_0 + c_1 t + a_0 \frac{t^2}{2 \times 1} + a_1 \frac{t^3}{3 \times 2} + \cdots + a_n \frac{t^{n+2}}{(n+2) \times (n+1)},$$

where c_0 and c_1 are arbitrary constants.

(ii) Solve $y'''(t) = P(t)$.

What happens if we throw a ball, rather than drop it? The position of the ball is now given by $\big(x(t), y(t)\big)$, where $y(t)$ is the height and $x(t)$ is the horizontal distance travelled. Galileo's experiments led him to the conclusion that the two coordinates could be treated independently, giving

$$x''(t) = 0,$$
$$y''(t) = -g.$$

We already know how to solve these equations to give

$$x(t) = x(0) + x'(0)t,$$
$$y(t) = y(0) + y'(0)t - \frac{gt^2}{2}.$$

Since the values of $x(0)$ and $y(0)$ only depend on our choice of origin, we may as well take $x(0) = y(0) = 0$ to obtain the simpler form

$$x(t) = ut,$$
$$y(t) = vt - \frac{gt^2}{2},$$

where $u = x'(0)$, $v = y'(0)$.

If we work to first order, we see that, at time δt, the particle is at $(x'(0)\delta t, y'(0)\delta t)$ and has travelled a distance

$$(x'(0)^2 + y'(0)^2)^{1/2} \delta t + o(\delta t) = (u^2 + v^2)^{1/2} \delta t + o(\delta t),$$

so it makes sense to say that the particle has initial speed $V = (u^2 + v^2)^{1/2}$. We write

$$u = V \cos\theta, \quad v = V \sin\theta.$$

Exercise 4.1.3. *Write* $\tan\theta$ *in terms of u and v.*

In gunnery, V is called the muzzle velocity and θ the angle of elevation. Writing our equations in the new notation, we get

$$x(t) = Vt\cos\theta,$$

$$y(t) = Vt\sin\theta - \frac{gt^2}{2}.$$

Thus $t = x(t)/(V\cos\theta)$ and

$$y(t) = x(t)\tan\theta - \frac{gx(t)^2(\sec\theta)^2}{2V^2}.$$

Removing the reference to t, we see that the *path* of the particle is given by

$$y = x\tan\theta - \frac{gx^2(\sec\theta)^2}{2V^2}. \qquad \bigstar\bigstar$$

Exercise 4.1.4. *Check the calculations just made and show that*

$$y = -\left(\frac{g^{1/2}\sec\theta}{2^{1/2}V}x - \frac{V\sin\theta}{2^{1/2}g^{1/2}}\right)^2 + \frac{V^2(\sin\theta)^2}{2g}.$$

Use this equation to show that, if $V\sin\theta > 0$, *the highest point on the trajectory is* $V^2(\sin\theta)^2/(2g)$ *above its initial height. How far has the projectile travelled horizontally when it reaches this point? Why did we impose the condition* $V\sin\theta > 0$?

Exercise 4.1.5. *If* $y = Y - B$, $x = X - A$ *and* $Y = CX^2$ *with* $C \neq 0$, *show that* $y = C(x + A)^2 + B$. *Explain why equation* $\bigstar\bigstar$ *tells us that the path of our projectile is indeed a parabola.*

If we throw a ball or fire a cannon, one of the things we wish to know is where the projectile will land, that is to say, we wish to find the values of x when $y = 0$. Substituting in equation $\bigstar\bigstar$, we get

$$0 = x\tan\theta - \frac{gx^2(\sec\theta)^2}{2V^2}$$

so that either $x = 0$ (corresponding to the point at which we start) or

$$\tan\theta - \frac{gx(\sec\theta)^2}{2V^2} = 0,$$

so

$$x = \frac{2V^2}{g}\cos\theta\sin\theta.$$

and, using a double angle formula (see Exercise 1.5.3 (iv)),

$$x = \frac{V^2}{g} \sin 2\theta.$$

We observe, at once, that the maximum range occurs when $\theta = \pi/4$ and that the projectile hits the ground at the same point if the angle of elevation is $\frac{\pi}{4} - \theta$ or $\frac{\pi}{4} + \theta$.

Galileo proves all this using geometry and we may surely echo one of his discussants.

> The force of rigid demonstration such as occur only in mathematics fills me with wonder and delight. From accounts given by gunners, I was already aware of the fact that in the use of cannons and mortars, the maximum range, that is the one in which the shot goes furthest, is obtained when the angle of elevation is 45 degrees ... but to understand why this happens far outweighs the mere information obtained by the testament of others or even by repeated experiment.
>
> *Discourses Relating to Two New Sciences*

4.2 Air resistance

Before proving the results of the previous section, Galileo explains that

> No matter how heavy the body, if it falls from a very considerable height, the resistance of the air will be such as to prevent any increase in speed and will render the motion uniform; and, in proportion as the moving body is less dense, this uniformity will be so much the more quickly attained and after a shorter fall. Even horizontal motion which, if no impediment were offered, would be uniform and constant is altered by the resistance of the air and finally ceases; and here again the less dense the body the quicker the process. Of the properties of [mass], of velocity and also of form, infinite in number, it is not possible to give any exact description; hence in order to handle the matter in a scientific way, it is necessary to cut loose from these difficulties; and having demonstrated these theorems in the case of no resistance, to use and apply them with such limitations as experience will teach.
>
> *Discourses Relating to Two New Sciences*

If we consider a body falling through air rather than a vacuum, it is plausible to replace the equation

$$y''(t) = -g$$

by

$$y''(t) = -g - h\big(v(t)\big),$$

where $v(t) = y'(t)$ and $h(v)$ is the 'resistance to motion when the body travels at velocity v'. We know that, as Galileo says, the function h depends in a complicated manner on the mass, shape and velocity of the body. However, we expect, in keeping with the ideas of the calculus, that

$$h(v) = h(0) + h'(0)v + o(v)$$

when v is small. We expect $h(0) = 0$ and so

$$h(v) = kv + o(v),$$

where k depends in a complicated manner on the mass and shape of the body.

Thus, provided that we remember that the approximation will break down when $y'(t)$ is large,[6] we expect the position of the body to satisfy

$$y''(t) = -g - ky'(t)$$

to a reasonable degree of accuracy.

As before, we seek to solve the differential equation in two stages by first considering the simpler equation

$$v'(t) = -g - kv(t). \qquad \bigstar$$

We will need the exponential function discussed in Section 3.2.

Exercise 4.2.1. *(Easy, but informative.) If $u(t) = A\exp(-kt)$, show that $u'(t) = -ku(t)$ and, writing $w(t) = (\exp kt)u(t)$, we have $w'(t) = 0$.*

Thought and (mathematical) experiment suggest looking at the function

$$f(t) = \exp(kt)\big(v(t) + g/k\big).$$

We observe that

$$\begin{aligned}
f'(t) &= k\exp(kt)\big(v(t) + g/k\big) + \exp(kt)v'(t) \\
&= \exp(kt)\big(kv(t) + g + v'(t)\big) = \exp(kt) \times 0 = 0,
\end{aligned}$$

so $f(t)$ is constant and

$$\exp(kt)\big(v(t) + g/k\big) = A$$

for some constant A.

Rearranging, we get

$$v(t) = -\frac{g}{k} + A\exp(-kt),$$

---

[6] And that we have no idea 'how large is large'.

that is to say, $y'(t) = -g/k + A \exp(-kt)$. Setting $t = 0$, we see that $A = V + g/k$, where $V = y'(0)$ is the initial velocity of the body. Thus

$$y'(t) = -\frac{g}{k} + \left(V + \frac{g}{k}\right)\exp(-kt). \qquad\qquad \bigstar\bigstar$$

The reader will not be surprised that we now consider

$$F(t) = y(t) + \frac{gt}{k} + \frac{1}{k}\left(V + \frac{g}{k}\right)\exp(-kt),$$

observe that $F'(t) = 0$ and deduce that

$$y(t) + \frac{gt}{k} + \frac{1}{k}\left(V + \frac{g}{k}\right)\exp(-kt) = B$$

for some constant B. Setting $t = 0$, we see that

$$y(0) + \frac{1}{k}\left(V + \frac{g}{k}\right) = B$$

and so

$$y(t) = y(0) - \frac{gt}{k} + \frac{1}{k}\left(V + \frac{g}{k}\right)\left(1 - \exp(-kt)\right).$$

Note that equation $\bigstar\bigstar$ tells us that, whatever speed it starts at, after a long time, the body will essentially have constant velocity $-g/k$.

Exercise 4.2.2. *The idea that we have just used to solve our particular differential equation can be used to solve a wide class of differential equations.*

(i) *Suppose that a is a constant and*

$$u'(t) - au(t) = h(t),$$

where $h(t)$ is a well behaved function of t. If

$$f(t) = \exp(-at)u(t) - \int_0^t \exp(-ax)h(x)\,dx,$$

show that $f'(t) = 0$ and deduce that

$$u(t) = B \exp at + \exp at \int_0^t \exp(-ax)h(x)\,dx$$

for some constant B.

(ii) *We continue the discussion of (i). Let C be a constant. If $h(t) = C$ for all t and $a \neq 0$, show that*

$$u(t) = A \exp at - \frac{1}{a}$$

for some constant A. What is the equation for $u(t)$ if $a = 0$?

If $h(t) = C \exp bt$ for all t and $a \neq b$, show that

$$u(t) = B \exp at + \frac{C}{b-a} \exp bt$$

for some constant B.

If $h(t) = C \exp at$ for all t, show that

$$u(t) = B \exp at + Ct \exp at$$

for some constant B.

(iii) *Suppose that* $a \neq b$ *and*

$$(s-a)(s-b) = s^2 + ps + q.$$

If

$$u''(t) + pu'(t) + qu(t) = 0$$

and $v(t) = u'(t) - au(t)$, *show that* $v'(t) - bv(t) = 0$. *Use the previous parts of the question to find* $v(t)$ *and show that*

$$u(t) = A \exp at + B \exp bt$$

for some constants A and B.

(iv) *Use a similar method to show that the general solution of*

$$u''(t) - 2au'(t) + a^2 u(t) = 0$$

is

$$u(t) = (A + Bt) \exp at.$$

(v) *Let p and q be as in part (iii). Find the general solution of*

$$u''(t) + pu'(t) + qu(t) = 1.$$

(vi) *Suppose that a, b, c are all different and that*

$$(s-a)(s-b)(s-c) = s^3 + ps^2 + qs + r.$$

Find the general solution of

$$u'''(t) + pu''(t) + qu'(t) + ru(t) = 0.$$

4.3 A dose of reality

In the preceding section, we modelled the fall of an object in a resisting medium by

$$y''(t) = -g - ky'(t)$$

on the grounds that, when the velocity $v = y'(t)$ is small, the resistance to motion $h(v)$ is approximately linear. Unfortunately, we had no idea how small v had to be. Huygens and other successors of Galileo did what we have not done and performed actual experiments. These experiments showed that in most of the cases of interest, $h(v)$ behaves more like some multiple of v^2! For things like rifle bullets, the resistance increases even more rapidly. (However, if we consider very small objects like dust motes falling through air or larger objects like steel balls falling through treacle, the velocities are small enough for the linear approximation to be useful.) Can we study the motion of falling bodies by solving the differential equation

$$y''(t) = -g - h(y'(t))$$

for more and more complicated functions h?

The first thing to observe is that the question 'can we solve a certain differential equation?' is more subtle than it seems. When we started this book, the only specific functions we talked about were the polynomials, the trigonometric functions and functions constructed from them in various ways. The reader will readily accept that no such function will satisfy the differential equation

$$f'(x) = f(x), \quad f(0) = 1,$$

although the statement is too imprecise to admit of a formal proof.[7] We then defined the exponential function and discovered that it solved our differential equation. In Exercise 4.2.2, we discovered that we could use the exponential function to solve a wide range of differential equations of the form

$$f''(x) + pf'(x) + qf(x) = 0.$$

We should thus replace the question 'can we solve a certain differential equation?' by 'can we solve a certain differential equation in terms of a certain collection of functions?'. The experience of the past three centuries is that the answer will often be no, and we then need to answer a different set of questions.

---

[7] However, if the reader decides on a precise formulation, she may be able to show that exp increases too fast to be obtained in any natural manner using only polynomials and trigonometric functions.

The first two questions that a pure mathematician might ask are the following.

(1) Does the equation (with appropriate starting conditions) actually have a solution?
(2) If a solution exists, is it unique?

The reader may give the traditional reply 'Nature has no trouble solving her own differential equations.' There is a great deal of force in this riposte, but, of course, the differential equations we study are not 'Nature's equations' but 'The equations we wish to model Nature with'. Moreover, if we direct a stream of water onto a horizontal surface (just run a tap into a kitchen sink), we see two regions in which Nature has no trouble solving her own differential equations separated by a 'hydraulic jump' where Nature is clearly having difficulties. Lightning strikes provide a grander example of Nature's difficulties in solving her own equations.

Exercise 4.3.1. *Explain why the differential equation*

$$f'(x)^2 = -1 - x^2$$

has no solution.

Exercise 4.3.2. *Here is a traditional example of non-uniqueness. Check that the differential equation*

$$f'(t) = 3f(t)^{2/3}$$

has two solutions $f(t) = 0$ and $f(t) = t^3$ both satisfying the condition $f(0) = 0$. (Actually, it has many others, but they require more discussion to justify.)

The kind of calculus discussed in this book is not capable of answering questions (1) and (2). Fortunately, the new rigorous analysis shows that, for a wide range of differential equations, the answer to both questions is yes.

Merely knowing that an answer exists is not much help by itself. What additional information do we want? At first sight, it looks as though the question we need to ask is 'what does the solution look like for a particular starting condition?', but really what we want to know is 'what do the solutions look like for a range of starting conditions?'. Galileo showed that, in the absence of resistance, the path of a cannon ball is parabolic, whatever the initial elevation of the cannon and whatever the muzzle velocity. If we change the elevation or the muzzle velocity slightly then the path will only change slightly.

Looking more deeply, we see that what we wish to know is 'what do the solutions look like for a range of similar differential equations and a range of similar starting conditions?'. Let us see what we can do in the case of a falling body.

It will be convenient to reverse the direction of the y-axis so that $y(t) - y(0)$ now measures the distance *fallen*. (Note that this means that we replace $-g$ by g in our equations.) If we study a falling body by means of the differential equation

$$y''(t) = g - h\big(y'(t)\big)$$

the resistance $h(v)$ will only be known roughly. and we need to allow for our lack of knowledge.

We expect $h(v)$ to be continuous, unbounded and strictly increasing with $h(0) = 0$ and $h(v) = -h(-v)$. There will thus be a unique V such that

$$h(V) = g$$

and, if $v'(t) = g - h\big(v(t)\big)$, we will have

$$v'(t) > 0 \text{ if } v(t) < V,$$
$$v'(t) < 0 \text{ if } v(t) > V.$$

Thus, under very general conditions, whatever speed it starts at, after a long time the projectile will essentially have constant *terminal velocity* V (in the direction of fall).

Exercise 4.3.3. *Suppose that h is as above, a, $b > 0$ and $h(v) \geq g + a$ for $v \geq V + b$. If*

$$v'(t) = g - h\big(v(t)\big)$$

and $v(0) \geq V + b$, use the mean value inequality to show that

$$v(t) \leq V + b$$

for $t \geq (v(0) - V)/a$.

There are rare accounts of people who have fallen from aeroplanes onto snow-covered pine forests and survived. They owe their survival not only to their soft landing, but to the fact that, once they reached their terminal velocity V, the height from which they fell was irrelevant. The parachute should be viewed as a method of lowering the user's terminal velocity.

Exercise 4.3.4.• *As we shall see in Section 9.1, the equations for the flight of projectile under realistic laws of resistance are rather complicated and require a new approach. However, as a purely mathematical exercise, we shall look at* *the simplest case when the equations are*

$$x''(t) = -kx'(t)$$
$$y''(t) = -ky'(t) - g,$$

where g and k are constant, k, $g > 0$ and $x(0) = y(0) = 0$, $x'(0) = u_0 > 0$, $y'(0) = v_0$. We take the y-axis upwards.

(i)   Solve the equations.

(ii) Show that $x'(t) \approx 0$, $y'(t) \approx -g/k$ for t large. Show also that $x(t) \approx u_0/k$ when t is large, but $x_t < u_0/k$ whenever $t \geq 0$. Sketch the path of the projectile for $v_0 = u_0 > 0$. (You may think of the projectile as launched from the top of a high cliff.)

(iii) Show that the equation of the path is

$$y = \left(\frac{c + v_0}{u_0} \right) x - \frac{c}{k} \log \left(\frac{u_0}{u_0 - kx} \right),$$

where $c = g/k$.

Exercise 4.3.5. *Here is another application of the exponential function in describing the physical world.*

A simple model of the atmosphere looks at a thin vertical cylindrical column of air of cross section A. The pressure $P(x)$ at height x is proportional to the weight $W(x)$ of the column of air above it divided by A. The weight of a volume V of air at height x is proportional to the density and the density is proportional to the pressure. By considering $W(x + \delta x) - W(x)$, or otherwise, show that

$$W'(x) = -K P(x)$$

and thus

$$P'(x) = -c P(x)$$

for some constants K and c. By solving the differential equation, show that

$$P(x) = P(0) \exp(-cx).$$

[The formula fits the facts rather better than we might expect, but we have not taken into account things like the change of temperature with altitude.]

5

Compound interest and horse kicks

5.1 Compound interest

If I invest a sum of money M at interest $100a\%$ per year paid annually, then, at the end of the year, I will have a sum $M(1 + a)$. (For the moment we suppose $a > 0$.) If I reinvest the entire sum, I will have $M(1 + a)^2$ at the end of the second year. If I continue in this way, I will have $M(1 + a)^N$ at the end of the Nth year. We say that the sum is invested at compound interest paid yearly.

If I invest the same sum at the same rate of interest, but paid every half year, then, at the end of six months, I will have $M(1 + a/2)$ and, at the end of a year, I will have $M(1 + a/2)^2$. Since

$$M(1 + a/2)^2 = M(1 + a + a^2/4) > M(1 + a),$$

I am better off with the new system.[1]

Is it always true that decreasing the payment period increases the amount that I have at the end of the year, that is to say

$$M \left(1 + \frac{a}{N + 1}\right)^{N+1} \underset{?}{\geq} M \left(1 + \frac{a}{N}\right)^N .$$

or, dividing through by the superfluous M,

$$\left(1 + \frac{a}{N + 1}\right)^{N+1} \underset{?}{\geq} \left(1 + \frac{a}{N}\right)^N ?$$

Exercise 5.1.1. *Try proving this by algebra.*

---

[1] Not much better off if a is small, since then $a^2/4$ is a second order quantity, but, as Mr Micawber says, 'Annual income twenty pounds, annual expenditure nineteen nineteen six, result happiness. Annual income twenty pounds, annual expenditure twenty pounds ought and six, result misery.' (Dickens, *David Copperfield*.)

One way that we could approach the problem is to apply the logarithm function to both sides, so converting our problem to the proof (or disproof) of the relation

$$(N+1)\log\left(1+\frac{a}{N+1}\right) \underset{?}{\geq} N\log\left(1+\frac{a}{N}\right).$$

Having committed ourselves this far, it is reasonable to seek to demonstrate the stronger result that

$$f(x) = x\log\left(1+\frac{a}{x}\right)$$

increases as x increases (for $x \geq 1$).

Exercise 5.1.2. *Show that*

$$f(x) = x(\log(a+x) - \log x).$$

We know that one way to investigate whether a function is increasing is to look at its derivative. Differentiating, we obtain

$$f'(x) = \log(a+x) - \log x + x\left(\frac{1}{a+x} - \frac{1}{x}\right)$$
$$= \log(a+x) - \log x - \frac{a}{a+x}.$$

We do not seem much further advanced, but a further differentiation gives, as the reader should check,

$$f''(x) = \frac{1}{a+x} - \frac{1}{x} + \frac{a}{(a+x)^2} = -\frac{a^2}{x(a+x)^2} < 0$$

whenever $x \geq 1$.

Exercise 5.1.3. *Check the computations of $f'(x)$ and $f''(x)$.*

Since $f''(x) < 0$ for $x \geq 1$, we know that f' is decreasing. Thus, if $f'(x_0) = -u_0 < 0$ for some $x_0 \geq 1$, we will have

$$\log\left(1+\frac{a}{x}\right) - \frac{a}{a+x} = f'(x) \leq -u_0$$

whenever $x \geq x_0$ and this is clearly false when x is very large. Thus $f'(x) \geq 0$ whenever $x \geq 1$ and $f(x)$ is increasing for $x \geq 1$, as we hoped.

Exercise 5.1.4.● **[Depreciation]** *Suppose that $0 < b < 1$ and N is a strictly positive integer. Show that*

$$\left(1-\frac{b}{N+1}\right)^{N+1} \geq \left(1-\frac{b}{N}\right)^{N}.$$

Can we obtain an immense fortune by having interest paid every day or every minute? The answer is given by looking at the mean value inequality.

Exercise 5.1.5. *If $f(x) = \log(1 + x) - x$, compute $f'(x)$ and $f(0)$. If $|x| \leq 1/2$ show that $|f'(x)| \leq 2|x|$. Deduce, by applying the mean value inequality, that $|f(h)| \leq 2h^2$ for $|h| \leq 1/2$.*

Thus, if a is any constant,

$$\left| N \log\left(1 + \frac{a}{N}\right) - a \right| = N \left| \log\left(1 + \frac{a}{N}\right) - \frac{a}{N} \right| \leq N \times 2\left(\frac{a}{N}\right)^2 = \frac{2a^2}{N},$$

provided only that N is large enough. The mean value inequality also tells us that there exists an M depending on a such that

$$|\exp(a + k) - \exp a| \leq M|k|$$

for $|k| \leq 1$. (This result does not depend on the sign of a, but, in our general discussion, we shall continue to suppose that $a > 0$.)

Exercise 5.1.6. *Show that, if $a > 0$, we can take $M = \exp(a + 1)$.*

Putting our two results together, we get

$$\left| \left(1 + \frac{a}{N}\right)^N - \exp a \right| = \left| \exp\left(N \log\left(1 + \frac{a}{N}\right)\right) - \exp a \right|$$
$$\leq M \frac{2a^2}{N} = \frac{K}{N}$$

for an appropriate constant K, provided only that N is large enough. Note that M and K will change if we change a.

Thus, however often we get our interest paid, we cannot do better at the end of the year than multiply our initial investment by $\exp a$.

Exercise 5.1.7. *(i) Why do institutions which borrow from the general public at low interest rarely compound at intervals shorter than a year? Why do institutions which lend at high interest[2] often compound at intervals shorter than a year?*

(ii) If I invest at c% per annum compounded yearly, how many years will it take to double my capital? Show that if c is small the answer is roughly $69/c$ years. (For $c = 3\%$ and $c = 4\%$ the 'rule of 72', where we replace 69 by 72, is closer.)

---

[2] Very high interest rates used to be associated with criminal enterprises, but, in the UK, you can now see TV advertisements for firms who will lend you £400 for 30 days and only require you to pay back £585.

(iii) If an institution invests 1 *dollar at* 3% *per annum compounded yearly, how much will it have after* 2000 *years? Where is the catch? (Actually there are lots of catches.)*

5.2 Digging tunnels

There is another way of looking at interest when we compound over very short periods. This is to look at the amount $f(t)$ we owe at time t and observe that, at the end of a short period δt, we incur further interest of about $b \times f(t) \times \delta t$ (where the nominal annual interest is $100b\%$). Thus

$$f(t + \delta t) = f(t) + bf(t)\delta t + o(\delta t).$$

Without worrying too much about 'how short is short', it is reasonable to interpret this as

$$f'(t) = bf(t)$$

from which we deduce, in the manner of Section 4.2, that

$$f(t) = f(0)\exp bt.$$

We can extend this idea further. Suppose that we wish to dig a long tunnel using borrowed money. If we owe $F(t)$ at time t, then, in the further short time δt, we incur a further interest of about $b \times F(t) \times \delta t$ (where the nominal annual interest is $100b\%$) and we must borrow a further $w\delta t$ to pay wages and other costs. Thus

$$F(t + \delta t) = F(t) + (bF(t) + w)\delta t + o(\delta t)$$

and

$$F'(t) = bF(t) + w.$$

We have already solved such an equation when we discussed falling bodies in Section 4.2 and we know that it yields

$$F(t) = -\frac{w}{b} + A\exp bt$$

for some constant A.

Exercise 5.2.1. *Redo the calculation just referred to.*

When we started the tunnel at time $t = 0$, we needed to buy equipment and set up a base at a cost c, say, so

$$c = F(0) = -\frac{w}{b} + A$$

and $A = c + (w/b)$. Thus $F(t) = (c + w/b)\exp(bt) - (w/b)$. In particular, if the tunnel is completed at time T, we see that we will then owe

$$L(b) = (c + w/b)\exp(bT) - (w/b). \qquad\qquad \bigstar$$

The nature of the problem tells us that b, c and w are all strictly positive.

Exercise 5.2.2. *(i) Show by differentiation that*

$$\frac{\exp x - 1}{x}$$

is an increasing function of x for $x > 0$ and deduce that, if $k > 0$,

$$\frac{\exp kx - 1}{x}$$

is an increasing function of x for $x > 0$.

(ii) If $L(b) = (c + w/b)\exp(bT) - (w/b)$ (with c and w fixed and strictly positive) show, by using (i) or otherwise, that L is an increasing function of b. Why should we expect this?

Once the tunnel is completed, we can draw an income from it, but we still have to pay interest. If our income is u per unit time, it is clear that we will eventually pay back our debts if $u > bL(b)$, but we will get further and further into debt if $u < bL(b)$.

Exercise 5.2.3. *Suppose that we owe a sum L at time $t = 0$, that our income is u per unit time (so between the time t and the time $t + s$ we receive su), the nominal annual interest is $100b\%$ and we pay off our debts as quickly as possible. Find a differential equation for $h(t)$, the amount we owe at time t (until we pay off our debts), and solve it. If $u > bL$, show that we will pay off our debt in time*

$$\frac{1}{b}\log\left(\frac{u}{u - bL}\right).$$

Using $\bigstar$ and the discussion preceding Exercise 5.2.3, we see that the condition that we should eventually pay off our debts is that

$$(bc + w)\exp(bT) - w < u.$$

Because exp is a rapidly increasing function, the result is very sensitive to changes in b and T. This explains why such projects are so vulnerable to delays and changes in interest rates.

What happens if we start digging the tunnel from both ends? This is equivalent to digging two tunnels with the same initial costs and wage costs per unit time as our original tunnel, but will only take half the time. Thus we will owe

$$2\big((c + w/b)\exp(bT/2) - w/b\big)$$

at the moment when the tunnel is completed and we will be better off provided that

$$(c + w/b)\exp(bT) - (w/b) > 2\big((c + w/b)\exp(bT/2) - (w/b)\big),$$

that is to say,

$$(c + w/b)\big(\exp(bT) - 2\exp(bT/2)\big) > -w/b.$$

Exercise 5.2.4. *(i) Show this condition can be rewritten as*

$$\big(\exp(bT/2) - 1\big)^2 > \frac{cb}{cb + w}.$$

(ii) Show that, if $c = 0$, we will always be better off starting from both ends. Why should we expect this?

(iii) Show that, whatever the value of c, we will be better off starting from both ends if

$$T > \frac{2}{b}\log 2.$$

Why should we expect there to be a T_0 (depending on b) such that, if $T > T_0$, we should always dig from both ends?

Thus, if interest rates are high and the project will take a long time, it is better to dig from both ends. For some projects it may be possible and desirable to start in several places at once.

When we discussed projectiles, our models were very close to reality and 'verbal reasoning' could not replace 'mathematical reasoning'. The discussion in this section deals with a very simplified model of a very complicated situation. Many people would say that all we have done is to provide a spuriously exact version of an argument that could have been obtained without any mathematics. The reader must decide for herself if she has learned anything from our way of proceeding.

5.3 Horse kicks

The rest of this chapter involves some simple probability. Even if the reader lacks the background to follow all of the arguments, I hope that she will find the general ideas helpful.

Let $n \geq r \geq 1$. Suppose that we have r *distinguishable* balls to place in n holes. We can place the first ball in any of the of the n holes, the second in any of $n - 1$ holes, the third in any of $n - 2$ holes and so on until we have placed the rth in one of $n - r + 1$ holes. There are thus

$$n \times (n - 1) \times (n - 2) \times \cdots \times (n - r + 1)$$

different ways of placing r *distinguishable* balls in n holes. The balls occupy r holes and could have been placed in those r holes in

$$r \times (r - 1) \times (r - 2) \times \cdots \times 1$$

ways. Thus there are

$$\frac{n \times (n - 1) \times (n - 2) \times \cdots \times (n - r + 1)}{r \times (r - 1) \times (r - 2) \times \cdots \times 1} = \binom{n}{r}$$

ways in which we can place r *indistinguishable* balls in n holes. (If you have not met $\binom{n}{r}$, pronounced 'n choose r' before, you can take the previous equation as a *definition*.)

Exercise 5.3.1. *Complete the following table, showing the different ways in which two balls labelled 1 and 2 can be placed in 4 holes.*

$$
\begin{array}{cccc}
1 & 2 & * & * \\
1 & * & 2 & * \\
1 & * & * & 2 \\
* & 1 & 2 & * \\
\vdots & \vdots & \vdots & \vdots
\end{array}
$$

Complete the following table, showing the different ways in which two blue balls can be placed in 4 holes.

$$
\begin{array}{cccc}
B & B & * & * \\
B & * & B & * \\
B & * & * & B \\
* & B & B & * \\
\vdots & \vdots & \vdots & \vdots
\end{array}
$$

Check that our general argument applies.

Now suppose that we toss a coin which has probability p of coming down heads and probability $1 - p$ of coming down tails. If we toss the coin n times, then the probability of a *particular* sequence of r heads and $n - r$ tails is $p^r(1 - p)^{n-r}$ and, since there are $\binom{n}{r}$ such sequences, the probability, q_r, of *some* sequence of r heads and $n - r$ tails is given by

$$q_r = \binom{n}{r} p^r (1 - p)^{n-r}.$$

Exercise 5.3.2. We make the definition $\binom{n}{0} = 1$. Check that the formula just given remains valid when $r = 0$.

If we repeat the process many times, we might expect that, on average, we will get about

$$\mathcal{E} = 0 \times q_0 + 1 \times q_1 + 2 \times q_2 + \cdots + n \times q_n$$

heads per attempt.[3] The reader may prefer to be more cautious and simply accept that $\mathcal{E}$ (which we refer to as the *expected number of heads*) is a useful quantity associated with the process.

We have

$$\mathcal{E} = 0 \times \binom{n}{0}(1 - p)^n + 1 \times \binom{n}{1} p(1 - p)^{n-1} + 2 \times \binom{n}{2} p^2 (1 - p)^{n-2}$$
$$+ \cdots + n \times \binom{n}{n} p^n,$$

but how do we compute it? The next couple of exercises give an amusing answer.

Exercise 5.3.3. [The binomial theorem] By considering the number of ways that we can choose either an x (a blue painted ball) or a y (a red painted ball) from a n bracketed pairs $(x + y)$ (sacks containing one red and one blue ball) in such a way that we take $n - r$ copies of x and r copies of y, show that

$$(x + y)^n = \binom{n}{0} x^n + \binom{n}{1} x^{n-1} y + \binom{n}{2} x^{n-2} y^2 + \cdots + \binom{n}{n} y^n.$$

---

[3] I apologise for deviating from the policy announced in *A shortage of letters* on page 3. $\mathcal{E}$ is pronounced 'curly E'.

By taking particular values of x and y, show that

$$\binom{n}{0} + \binom{n}{1}y + \binom{n}{2}y^2 + \cdots + \binom{n}{n}y^n = (1+y)^n,$$

$$\binom{n}{0} + \binom{n}{1} + \binom{n}{2} + \cdots + \binom{n}{n} = 2^n,$$

and

$$\binom{n}{0} - \binom{n}{1} + \binom{n}{2} - \cdots + (-1)^n\binom{n}{n} = 0.$$

Exercise 5.3.4. *By differentiating both sides of the binomial expansion*

$$(1+x)^n = \binom{n}{0} + \binom{n}{1}x + \binom{n}{2}x^2 + \cdots + \binom{n}{n}x^n,$$

show that

$$n(1+x)^{n-1} = 0 \times \binom{n}{0} + 1 \times \binom{n}{1} + 2 \times \binom{n}{2}x + \cdots + n \times \binom{n}{n}x^{n-1}.$$

By setting $x = p/(1-p)$ and simplifying, show that

$$\mathcal{E} = np.$$

Now suppose that p is very small (so that the event of a head is very rare), but n is very large (so we have many trials). More precisely, let us look what happens when we set $p = \mathcal{E}/n$ (so the expected number of heads is fixed) and make n large. Then the probability of r heads is

$$
\begin{aligned}
q_r &= \binom{n}{r}p^r(1-p)^{n-r} \\
&= \frac{n \times (n-1) \times \cdots (n-r+1)}{r!}p^r(1-p)^{n-r} \\
&= \frac{n \times (n-1) \times \cdots (n-r+1)}{r!}\left(\frac{\mathcal{E}}{n}\right)^r\left(1-\frac{\mathcal{E}}{n}\right)^{n-r} \\
&= 1 \times \left(1-\frac{1}{n}\right) \times \left(1-\frac{2}{n}\right) \times \cdots \times \left(1-\frac{r-1}{n}\right) \times \left(1-\frac{\mathcal{E}}{n}\right)^{-r} \\
&\quad \times \frac{1}{r!}\left(1-\frac{\mathcal{E}}{n}\right)^n \mathcal{E}^r.
\end{aligned}
$$

If we now observe that

$$1 \times \left(1-\frac{1}{n}\right) \times \left(1-\frac{2}{n}\right) \cdots \times \left(1-\frac{r-1}{n}\right) \times \left(1-\frac{\mathcal{E}}{n}\right)^{-r} \approx 1,$$

when n is large compared with r, and remember the approximation we found in Section 5.1, we see that

$$q_r \approx \frac{\mathcal{E}^r}{r!} \exp(-\mathcal{E}) \qquad \bigstar$$

when n is sufficiently large.

What is true for large numbers of coin tosses with low probability of heads is presumably true for any event which has low probability of occurring in any particular case, but where the number of cases where it could occur is very large. Bortkiewicz analysed a number of such processes including the number of quadruplets born in Prussia in a given year and the number of deaths from horse kicks in the Prussian cavalry in a given year. In each case the results correspond to the *Poisson distribution* given by equation $\bigstar$.

5.4 Gremlins

Cars and people have a greater tendency to break down when they are very new or very old. However, some types of electrical components have a break-down rate which is independent of age.

Let us try to model such a piece of equipment by supposing that it is in the charge of a gremlin.[4] The gremlin tosses a coin every $1/n$th minute and each toss has a probability $\mathcal{E}/n$ of landing heads. When the coin lands heads, the gremlin causes the component to fail. The probability that the component has not failed after T minutes is

$$\left(1 - \frac{\mathcal{E}}{n}\right)^{Tn} = \left(\left(1 - \frac{\mathcal{E}}{n}\right)^n\right)^T \approx \exp(-T\mathcal{E}).$$

Now suppose that the component is replaced immediately when it fails. The probability, $p_r(T)$, that it is replaced exactly r times in T minutes is the probability that our gremlin throws r heads in Tn tosses. Thus

$$p_r(T) = \binom{Tn}{r}\left(\frac{\mathcal{E}}{n}\right)^r\left(1 - \frac{\mathcal{E}}{n}\right)^{Tn-r} = \binom{Tn}{r}\left(\frac{T\mathcal{E}}{Tn}\right)^r\left(1 - \frac{T\mathcal{E}}{Tn}\right)^{Tn-r}.$$

The same argument that we used for horse kicks shows that

$$p_r(T) \approx \frac{(T\mathcal{E})^r}{r!} \exp(-T\mathcal{E}).$$

---

[4] According to the Oxford Dictionary, a gremlin is a 'mischievous sprite imagined as the cause of mishaps to aircraft; later, an embodiment of mischance in other activities.'

The process we are describing has a further 'forgetfulness property'. Suppose that Anthony has been watching the gremlin for many hours, but Bertha has just begun. Since the past history of coin tossing has no effect on future tosses, Anthony will be no better at predicting the time of the next break down than Bertha. Thus, for example, however long the system has run without failure, the expected time (that is to say the average time) to the next failure remains unchanged.

Nature does not throw a coin every $1/n$ minutes, but, observing that our approximations get better and better as we make n bigger, we expect there to be processes such that 'probability an event occurs in time δt is $\mathcal{E}\delta t + o(\delta t)$ independent of previous history'. Our arguments strongly suggest that the probability that no such event happens in time T will be $\exp(-\mathcal{E}T)$ and that $p_r(T)$, the probability that exactly r events occur in time T will be given by

$$p_r(T) = \frac{(T\mathcal{E})^r}{r!} \exp(-T\mathcal{E}).$$

Examples of such systems range from calls to a call centre (where the event is the reception of a new call) to the clicks of a Geiger counter (where the event is a new click) and the failure of neon bulbs.

6
Taylor's theorem

6.1 Do the higher derivatives exist?

Mathematicians sometimes express opinions about non-mathematical topics to the intense annoyance of those who feel that they should stick to their own business. In 1734, a very annoyed Bishop Berkeley decided to attack mathematicians on their home ground and wrote a short, but extremely clever, pamphlet entitled

THE ANALYST
A Discourse addressed to an Infidel Mathematician.
Wherein it is examined
whether the Object, Principles, and Inferences
of the modern Analysis are more distinctly conceived,
or more evidently deduced, than Religious Mysteries and Points of Faith.

In it, he argues that the calculus as then conceived was such a tissue of unfounded assumptions as to remove every shred of authority from its practitioners. If the reader substitutes our 'derivative' for words like 'fluxions', 'differences' and 'infinitesimals' she will get the flavour of his attack

> And yet in the calculus differentialis, which Method serves to all the same Intents and Ends with that of Fluxions, our modern Analysts are not content to consider only the Differences of finite Quantities: they also consider the Differences of those Differences, and the Differences of the Differences of the first Differences. And so on ad infinitum. That is, they consider Quantities infinitely less than the least discernible Quantity; and others infinitely less than those infinitely small ones; and still others infinitely less than the preceding Infinitesimals, and so on without end or limit. Insomuch that we are to admit an infinite succession of Infinitesimals, each infinitely less than the foregoing, and infinitely greater than the following. As there are first, second, third, fourth, fifth, etc. Fluxions, so there are Differences, first, second, third, fourth, etc. in an infinite Progression towards nothing, which you still approach and never arrive at . . .

95

It must indeed be acknowledged, the modern Mathematicians do not consider
these Points as Mysteries, but as clearly conceived and mastered by their
comprehensive Minds. They scruple not to say, that by the help of these new
Analytics they can penetrate into Infinity itself: That they can even extend their
Views beyond Infinity: that their Art comprehends not only Infinite, but Infinite of
Infinite (as they express it) or an Infinity of Infinites. But, notwithstanding all these
Assertions and Pretensions, it may be justly questioned whether, as other Men in
other Inquiries are often deceived by Words or Terms, so they likewise are not
wonderfully deceived and deluded by their own peculiar Signs, Symbols, or
Species. Nothing is easier than to devise Expressions or Notations for Fluxions and
Infinitesimals of the first, second, third, fourth, and subsequent Orders, proceeding
in the same regular form without end or limit $\dot{x}$, $\ddot{x}$, $\dddot{x}$, $\ddddot{x}$ etc. [that is to say, our
$f'(t)$, $f''(t)$, $f'''(t)$, $f''''(t)$, etc.] .... These Expressions indeed are clear and
distinct, and the Mind finds no difficulty in conceiving them to be continued
beyond any assignable Bounds. But if we remove the Veil and look underneath, if
laying aside the Expressions we set ourselves attentively to consider the things
themselves, which are supposed to be expressed or marked thereby, we shall
discover much Emptiness, Darkness, and Confusion.

On the same theme, he says 'the Velocities of the Velocities, the second,
third, fourth, and fifth Velocities, etc. exceed, if I mistake not, all Humane
Understanding. ... He who can digest a second or third Fluxion, a second or
third Difference, need not, methinks, be squeamish about any Point in Divinity.'

The sting of Berkeley's attack has been removed by the development of
rigorous analysis. However, it is worth taking a little while to think about
Berkeley's remarks about higher derivatives.

From the *mathematical* point of view there is no problem. Starting from a
well behaved function f we can form its derivative f' *which is just another
function*. If everything is well behaved, we can differentiate the new function
f' to obtain f'' and then differentiate that to obtain f''' and so on. (The result
of differentiating n times is called the nth derivative and written $f^{(n)}$.)

From the point of view of intuition, things are less clear. I think that riding
in cars, trains and aeroplanes has given me a clear experience of velocity and
some experience of acceleration. Thus, I believe (possibly falsely) that I have
a certain intuitive idea of a rate of change and some (rather less clear) intuitive
idea of the rate of change of a rate of change. However, the notion of the rate
of change of a rate of change of a rate of change conveys rather little to me.[1]

To see that this is not entirely a consequence of my mental limitations, it
is instructive to consider the problem of physical measurement. If I ask an

---

[1] 'In the fall of 1972, President Nixon announced that the rate of increase of inflation was
decreasing. This was the first time a sitting president had used the third derivative to advance
his case for re-election.' Hugo Rossi in the *Notices of the AMS*, Vol. 43, No. 10, Oct. 1996.

engineer to measure a quantity (distance travelled, say), then she will often be able to do this easily and cheaply. If I ask for the rate of change of the quantity (velocity, say), then the measurement will considerably harder to perform and cost considerably more. If I go further and ask for the rate of change of the rate of change (acceleration, say), the task becomes very much harder and very much more expensive.

The next exercise may help the reader see part of the problem.

Exercise 6.1.1. *Let*

$$f(x) = \sin x + 10^{-20} \sin 10^{12}x.$$

Graph f, f' and f'' using appropriate scales.

Ever since the work of Galileo on falling bodies, we have known that, hard though second derivatives may be to measure and understand, they form part of the language of nature. Fortunately (either because that is the way nature is, or because that is the way we understand nature) many of our physical theories do not require higher derivatives than these. We may answer Berkeley by saying that we can differentiate as many times as we like, but that we do not expect to have much intuition as to the meaning of derivatives of high order.

6.2 Taylor's theorem

In Section 1.4 we considered the linear approximation

$$f(x + h) = f(x) + f'(x)h + o(h)$$

and I claimed that quadratic and higher approximations could be obtained without using any further tools. The time has come to justify my claim.

Suppose first that $a > 0$ and g is a well behaved function with

$$g(0) = g'(0) = g''(0) = \cdots = g^{(n-1)}(0) = 0$$

and $|g^{(n)}(t)| \le M$ whenever $0 \le t \le a$.

Since $-M \le g^{(n)}(t) \le M$, the inequality rule for integration (see page 35) shows that

$$\int_0^x (-M) \, dt \le \int_0^x g^{(n)}(t) \, dt \le \int_0^x M \, dt$$

and so, using the fundamental theorem of the calculus,

$$-Mx \le g^{(n-1)}(x) - g^{(n-1)}(0) \le Mx.$$

Recalling that $g^{(n-1)}(0) = 0$ and renaming the variable, we obtain

$$-Mt \leq g^{(n-1)}(t) \leq Mt$$

whenever $0 \leq t \leq a$.

If doing something once appears to produce information, it is always worth trying it again. The inequality rule for integration shows that

$$\int_0^x (-Mt)\, dt \leq \int_0^x g^{(n-1)}(t)\, dt \leq \int_0^x Mt\, dt$$

and so, using the fundamental theorem of the calculus,

$$-M\frac{x^2}{2} \leq g^{(n-2)}(x) - g^{(n-2)}(0) \leq M\frac{x^2}{2}.$$

Recalling that $g^{(n-2)}(0) = 0$ and renaming the variable, we obtain

$$-M\frac{t^2}{2 \times 1} \leq g^{(n-2)}(t) \leq M\frac{t^2}{2 \times 1}$$

whenever $0 \leq t \leq a$.

We repeat the argument again to obtain

$$-M\frac{t^3}{3 \times 2 \times 1} \leq g^{(n-3)}(t) \leq M\frac{t^3}{3 \times 2 \times 1}$$

whenever $0 \leq t \leq a$. Continuing in this way, we eventually get to the formula

$$-M\frac{t^n}{n \times (n-1) \times \cdots \times 3 \times 2 \times 1} \leq g(t) \leq M\frac{t^n}{n \times (n-1) \times \cdots \times 3 \times 2 \times 1}$$

or, in more condensed notation,

$$|g(t)| \leq M\frac{t^n}{n!}$$

whenever $0 \leq t \leq a$.

Exercise 6.2.1. *If $h(t) = g(-t)$, show that $h'(t) = -g'(-t)$. Show, more generally, that $h^{(r)}(t) = (-1)^r g^{(r)}(-t)$. Deduce that, if we add the further condition $|g^{(n)}(t)| \leq M$ for $-a \leq t \leq 0$ to those placed on g in the previous discussion, then*

$$|g(t)| \leq M\frac{|t|^n}{n!}$$

whenever $-a \leq t \leq 0$.

We have shown that

$$|g(t)| \leq M\frac{|t|^n}{n!}$$

whenever $|t| \leq a$, provided that g satisfies some rather restrictive conditions. However, we can easily extend our result to more general functions.

Suppose first that f is a well behaved function and $|f^{(n)}(t)| \leq M$ whenever $|t| \leq a$. If we write

$$g(t) = f(t) - f(0) - \frac{f'(0)}{1!}t - \frac{f''(0)}{2!}t^2 - \cdots - \frac{f^{(n-1)}(0)}{(n-1)!}t^{n-1},$$

then g obeys the conditions we have set out. (That is to say, g is a well behaved function with

$$g(0) = g'(0) = g''(0) = \cdots = g^{(n-1)}(0) = 0$$

and $|g^{(n)}(t)| \leq M$ whenever $|t| \leq a$.)

Exercise 6.2.2. *Verify the statement just made.*

Our earlier result on g now translates into a result on f:

$$\left| f(t) - f(0) - \frac{f'(0)}{1!}t - \frac{f''(0)}{2!}t^2 - \cdots - \frac{f^{(n-1)}(0)}{(n-1)!}t^{n-1} \right| \leq M \frac{|t|^n}{n!}$$

whenever $|t| \leq a$.

Although I shall follow custom[2] and just call this Taylor's theorem, the reader should always think of it as 'Taylor's theorem *with a remainder estimate*' (or 'Taylor's theorem *with an error estimate*').

Exercise 6.2.3. *Use a simple argument to obtain the following, apparently more general, result.*

Suppose that F is a well behaved function, y is a fixed number and $|F^{(n)}(t)| \leq M$ whenever $|t - y| \leq a$. Then

$$\left| F(t) - F(y) - \frac{F'(y)}{1!}(t - y) - \frac{F''(y)}{2!}(t - y)^2 - \cdots - \frac{F^{(n-1)}(y)}{(n-1)!}(t - y)^{n-1} \right|$$

$$\leq M \frac{|t - y|^n}{n!}$$

whenever $|t - y| \leq a$.

[*Some people refer to this result as* Taylor's theorem *and the case when $y = 0$ as* Maclaurin's theorem – *a distinction without a difference.*]

---

[2] It turns out that there is a whole family of Taylor's theorems in advanced calculus, so, when doing advanced work, you should always check which one is intended.

Since $Mh^n/n!$ decreases to zero faster than h^{n-1}, Taylor's theorem with an error estimate tells us that

$$F(y+h) = F(y) + \frac{F'(y)}{1!}h + \frac{F''(y)}{2!}h^2 + \cdots + \frac{F^{(n-1)}(y)}{(n-1)!}h^{n-1} + o(h^{n-1}).$$
$$\bigstar$$

As the reader will expect, mathematicians pronounce '$+o(h^{n-1})$' as 'plus little o of h^{n-1}' but I strongly recommend that the reader pronounces it as 'plus an error term which diminishes faster than the $n-1$ th power of h'. The result given in equation $\bigstar$ is called a *local Taylor's theorem* since it deals with *small* changes in the variables. It tells us that functions which are well behaved from the point of view of the calculus (that is to say, functions which can be differentiated many times) look *locally* like polynomials.

Exercise 6.2.4. *If g is a function such that*

$$g(h) = a_0 + a_1 h + a_2 h^2 + \cdots + a_{n-1} h^{n-1} + o(h^{n-1})$$

and

$$g(h) = b_0 + b_1 h + b_2 h^2 + \cdots + b_{n-1} h^{n-1} + o(h^{n-1}),$$

explain (without referring to Taylor series) why $a_j = b_j$ for $0 \le j \le n-1$.
Deduce that, if F satisfies $\bigstar$ and we find c_j with

$$F(y+h) = c_0 + c_1 h + c_2 h^2 + \cdots + c_{n-1} h^{n-1} + o(h^{n-1}),$$

then $c_j = F^{(j)}(y)/j!$ for $0 \le j \le n-1$.

A simple application of the local Taylor's theorem occurs when we study local maxima and minima. Suppose that f is a well behaved function. We have already observed that

$$f(a+h) = f(a) + f'(a)h + o(h)$$

so, if $f'(a) \ne 0$, then changes in f near a are dominated by the linear term $f'(a)h$. In particular, if $f'(a) \ne 0$, f cannot attain a maximum or a minimum at a.

Now suppose that $f'(a) = 0$. Taking an extra term in the Taylor expansion, we get

$$f(a+h) = f(a) + f'(a)h + f''(a)\frac{h^2}{2!} + o(h^2) = f(a) + f''(a)\frac{h^2}{2!} + o(h^2)$$

so, if $f''(a) \ne 0$, then changes in f near a are dominated by the quadratic term

$(f'(a)/2!)h^2$. In particular, if $f''(a) > 0$, then f has a minimum at a and, if $f''(a) < 0$, then f has a maximum at a.

If the reader requires (as she should) a more precise argument, she should observe that if

$$f(a + h) = f(a) + Ah^2 + o(h^2)$$

and $A > 0$, then the o notation tells us that we can find a $u > 0$ such that

$$|f(a + h) - (f(a) + Ah^2)| \le \frac{A}{2}h^2$$

whenever $|h| \le u$. We thus have

$$f(a + h) - (f(a) + Ah^2) \ge -\frac{A}{2}h^2$$

so

$$f(a + h) \ge f(a) + \frac{A}{2}h^2 \ge f(a)$$

whenever $|h| \le u$. Thus f attains a local minimum at a.

Exercise 6.2.5. *Consider the case when*

$$f(a + h) = f(a) + Ah^2 + o(h^2)$$

and $A < 0$.

These ideas are of considerable theoretical importance, but, in my view, study of the sign of $f'(t)$ for t close to a is often quicker and more informative than attempting to calculate $f''(a)$.

Exercise 6.2.6. *Suppose that f is well behaved and*

$$f'(a) = f''(a) = \cdots = f^{(r-1)}(a) = 0,$$

but $f^{(r)}(a) \ne 0$. Does f attain a local maximum at a? Does f attain a local minimum at a? (Your answer will depend on the sign of $f^{(r)}(a)$ and on whether r is odd or even.)
[Note that, if $r \ge 2$, this result has little practical importance since there is, in general, no reason to expect $f'(t)$ and $f''(t)$ to vanish at the same time t.]

6.3 Calculation with Taylor's theorem

In Section 3.2 we introduced a new function exp, but did not show how to calculate it. Taylor's theorem gives us a possible approach. Let $a > 0$. Writing

$E(x) = \exp x$, we see that, if $|E^{(n)}(t)| \leq M$ whenever $|t| \leq a$, then

$$\left| E(t) - E(0) - \frac{E'(0)}{1!}t - \frac{E''(0)}{2!}t^2 - \cdots - \frac{E^{(n-1)}(0)}{(n-1)!}t^{n-1} \right| \leq M\frac{|t|^n}{n!}$$

whenever $|t| \leq a$. Since $E'(t) = E(t)$, we have $E^{(r)}(t) = E(t)$ and we obtain the elegant statement that, if $|E(t)| \leq M$ whenever $|t| \leq a$, then

$$\left| E(t) - 1 - \frac{t}{1!} - \frac{t^2}{2!} - \cdots - \frac{t^{n-1}}{(n-1)!} \right| \leq M\frac{|t|^n}{n!}$$

whenever $|t| \leq a$.

Exercise 6.3.1. *Verify the statement just made.*

We know that E is increasing, so $E(a) \geq E(t)$ whenever $t \leq a$ and we may choose $M = E(a)$ to obtain

$$\left| E(t) - 1 - \frac{t}{1!} - \frac{t^2}{2!} - \cdots - \frac{1}{(n-1)!} \right| \leq E(a)\frac{|t|^n}{n!}$$

whenever $|t| \leq a$. Taking $t = a$ gives

$$\left| E(a) - 1 - \frac{a}{1!} - \frac{a^2}{2!} - \cdots - \frac{a^{n-1}}{(n-1)!} \right| \leq E(a)\frac{a^n}{n!},$$

so

$$\exp a \approx 1 + \frac{a}{1!} + \frac{a^2}{2!} + \cdots + \frac{a^{n-1}}{(n-1)!}$$

with an error less than $(\exp a)a^n/n!$. This error can be made as small as desired by taking n large enough.

Exercise 6.3.2. *In the calculations above we took $a \geq 0$. Show that, if $a \leq 0$,*

$$\left| \exp a - 1 - \frac{a}{1!} - \frac{a^2}{2!} - \cdots - \frac{a^{n-1}}{(n-1)!} \right| \leq \frac{|a|^n}{n!}.$$

It is easy to check that

$$\log 4 = \int_1^4 \frac{1}{t}\,dt \geq \int_1^2 \frac{1}{2}\,dt + \int_2^3 \frac{1}{3}\,dt + \int_3^4 \frac{1}{4}\,dt = \frac{1}{2} + \frac{1}{3} + \frac{1}{4} > 1,$$

so that $\exp 1 \le 4$ and

$$\left| \exp 1 - 1 - \frac{1}{1!} - \frac{1^2}{2!} - \cdots - \frac{1}{(n-1)!} \right| \le \frac{4}{n!}.$$

Let us perform the appropriate calculations when $n = 6$.

Calculation. We write $u_0 = 1$ and $u_r = u_{r-1}/r$, so that $u_r = 1/r!$ and

$$|\exp 1 - (u_0 + u_1 + u_2 + \cdots + u_5)| \le u_6.$$

Using the arithmetic functions[3] on my pocket calculator, I get

$$u_0 = 1$$
$$u_1 = 1$$
$$u_2 = .5$$
$$u_3 \approx 0.166\,67$$
$$u_4 \approx 0.041\,67$$
$$u_5 \approx 0.008\,33$$
$$u_6 \approx 0.001\,39$$
$$4u_6 \approx 0.005\,56$$

giving

$$e = \exp 1 \approx u_0 + u_1 + u_2 + \cdots + u_5 \approx 2.718\,06$$

with an error of less than about 0.0056.

Exercise 6.3.3. *We can do slightly better. Explain why the computation just given shows that* $\exp 1 < 2.8$ *and so the error in the formula is less than* 0.004.

Exercise 6.3.4. *(i) Carry out the same calculation as we set out above, but with* $n = 11$.
(ii) Carry out the same calculation, but with $a = 1/10$ *and* $n = 6$.
(iii) Carry out the same calculation, but with $a = -1/10$ *and* $n = 6$.

Exercise 6.3.5.• *This exercise improves the result given as Exercise 5.1.7 on page 86. By first estimating*

$$\left| \left(1 + \frac{a}{N} + \frac{a^2}{N^2}\right) - \exp(a/N) \right|$$

---

[3] That is to say, only $+$, $-$, $\times$ and $\div$.

and then

$$\left| \log \left(1 + \frac{a}{N} + \frac{a^2}{N^2} \right) - \frac{a}{N} \right|,$$

or otherwise, show that there exists an L (depending on a) with

$$\left| \left(1 + \frac{a}{N} + \frac{a^2}{2N^2} \right)^N - \exp a \right| \leq \frac{L}{N^2}$$

provided that N is large enough.

It looks as though Taylor's theorem is a good way of computing $\exp a$, but, as the reader may be coming to expect, things are not quite as simple as they look.

Exercise 6.3.6. *Try to calculate* $\exp 10$ *using our method.*

What has gone wrong? We seek to approximate $\exp 10$ using the formula

$$\exp 10 \approx v_0 + v_1 + v_2 + \cdots + v_N$$

with $v_r = 10^r / r!$. Observe that

$$\frac{v_r}{v_{r-1}} = \frac{10}{r}$$

and so v_r increases with r to a maximum when $r = 9$ and $r = 10$ and then decreases rather slowly until $r = 20$.

Exercise 6.3.7. *Verify that* $v_{10} \approx 2756$ *and* $v_{20} \approx 41$.

Thereafter, the value of v_r is more than halved every time r increases by 1, so things get better quite rapidly.

This means that matters are not quite as bad as at first appears. Since $0 \leq v_{r+1} \leq v_r / 2$ for $r \geq 20$ we have $v_r \leq v_{20} 2^{20-r}$ whenever $r \geq 20$ and so, if $M \geq 21$,

$$v_{21} + v_{22} + \cdots + v_M \leq v_{20} \left(\frac{1}{2} + \frac{1}{4} + \frac{1}{8} + \cdots + \frac{1}{2^{M-20}} \right)$$
$$= v_{20}(1 - 2^{20-M}) \leq v_{20}.$$

Now we know that

$$\exp 10 \approx (v_0 + v_1 + \cdots + v_{20}) + (v_{21} + v_{22} + \cdots + v_M)$$

with an error which can be made as small as desired by taking M large enough. Thus

$$\exp 10 \approx v_0 + v_1 + \cdots + v_{20}$$

with an error less than v_{20}.

Exercise 6.3.8. *Use this result to show that* $\exp 10 \approx 22\,000$ *correct to three significant figures (more precisely,* $22\,050 > \exp 10 > 21\,950$*).*

The reader may feel that $\exp 10$ is so large that it is not surprising that we need a large number of terms to calculate it. However, this is by no means the whole story.

Exercise 6.3.9. *Try and calculate* $\exp(-10)$ *using the Taylor series.*

A little thought reveals the foolish nature of the task set out in Exercise 6.3.9. We know that $\exp 10 \approx 22\,000$ and so $\exp(-10) = 1/\exp 10 \approx 1/22\,000$. If we want $\exp(-10)$ correct to three decimal places, then we already know that the answer is zero and further calculation is futile. Suppose, none the less, that we try to calculate $\exp(-10)$ to one significant figure using the Taylor series. We get successive estimates

$$s_0 = v_0, \; s_1 = v_0 - v_1, \; s_2 = v_0 - v_1 + v_2, \; \ldots$$

with $v_r = 10^r/r!$. Since the change between s_r and s_{r+1} has magnitude v_{r+1}, it is pretty clear that, unless v_{r+1} is smaller than the desired error, s_r will only be sufficiently close to the desired answer 'by accident'. Thus we need to use at least $N + 1$ terms, where $v_N \approx 1/22\,000$ and this N will be pretty large.

This is not the most unpleasant fact that we must face. Suppose that we calculate s_{2M+1} with $2M + 1 \geq N$. Then, in effect, we are performing the subtraction $w_M - \tilde{w}_M$, where

$$w_M = v_0 + v_2 + \cdots + v_{2M} \text{ and } \tilde{w}_M = v_1 + v_3 + \cdots + v_{2M+1},$$

that is to say, we are calculating a very small number as the difference of two very large numbers. This procedure makes as much sense as weighing a feather by first weighing a battle tank with the feather on top, then removing the feather, weighing the tank again and taking the difference.

Should this worry us? Those who believe that mathematics can be done on autopilot should be worried, but the rest of us can take comfort from the following reflections.

(1) The only things that look precisely like polynomials are polynomials. Calculus, at the level of this book, is the study of functions which look *locally* like polynomials and there is no reason to expect this behaviour to persist over long ranges. As we have remarked earlier, the exponential function is extremely unpolynomial-like in its behaviour over long ranges.
(2) The fact that it is difficult to calculate $\exp x$ in one way does not mean that it is hard to calculate it in another.

Exercise 6.3.10. *A notorious French Minister of Education once stated that 'Mathematics is being devalued in an almost inescapable way. From now on there are machines for computation. Likewise for drawing graphs.'*[4] *Suppose that your computer can work to 32 significant figures. Show that direct use of the Taylor's theorem in the manner above to calculate* $\exp(-100)$ *will produce nonsense.*

Exercise 6.3.11. *In this exercise we estimate* $\exp(-100)$. *Recall the estimate for* $e = \exp 1$ *you obtained in Exercise 6.3.4 (i). Using the arithmetic functions on your pocket calculator*[5] *estimate*

$$e^2, \ e^4, \ e^8, \ e^{16}, \ e^{32}, \ \text{and} \ e^{64}.$$

Now estimate

$$\exp 100 = e^{100} = e^{64} \times e^{32} \times e^4$$

and so estimate $\exp(-100)$ *to four significant figures.*
[*This question merely shows that the thing can be done, not that this is the best way of doing it.*]

Exercise 6.3.12. *(i) Write out the form of Taylor's theorem with remainder estimate for the functions* cos *and* sin.

(ii) Use (i) to compute $\sin(1/10)$ *to six decimal places showing that your answer is correct.*

(iii) Show that, in principle, you can compute $\sin x$ *to any number of significant terms for any* x *by taking sufficiently many terms. Explain briefly why this is not a good idea when* x *is large.*

(iv) Explain how, provided you can find a certain constant (to be identified) to very high accuracy, you could easily compute $\sin x$ *to six figure accuracy for all* $|x| < 10^6$.

---

[4] 'Living? Our servants will do that for us.' (Villiers de l'Isle-Adam, *Axël.*) There was a certain grim satisfaction in the French mathematical community when, in later life, the ex-minister became a climate crank railing against computer models.
[5] Life will be easier if your calculator uses scientific notation.

Exercise 6.3.13. *In* Surely You're Joking, Mr Feynman!, *Feynman recalls boasting that he could provide an estimate to within an accuracy of ten percent of the correct answer to any problem that anyone could state in ten seconds. Feynman beat off all challenges until his friend Olum asked him to estimate* $\tan 10^{100}$.

 (i) Why did this stump Feynman?
 (ii) Can you solve Olum's problem on a calculator? Give reasons.
 (iii) Can you solve Olum's problem on a computer?

[*The reader should ask herself part (iii) once a year, since her answer may change as her knowledge increases.*]

Exercise 6.3.14. *Let* $\sin_D$ *be the sine function when we measure angle in degrees. Write out the form of Taylor's theorem with remainder for the function* $\sin_D$. *Comment.*

Exercise 6.3.15. *Write out the form of Taylor's theorem with remainder estimate for the function* $f(x) = (x + y)^m$ *and deduce the binomial theorem (see Exercise 5.3.3).*

7

Approximations, good and bad

7.1 Find the root

As we have seen, it is one thing to define a number like exp 10 and another thing to calculate it. If we do decide to calculate a number, then we should ask ourselves the following questions.

(1) How accurate do we want the answer? Do we need a guarantee of accuracy?
(2) Are we going do to this type of calculation only once or are we going to do it many times? If we are only going to do it once, then it is probably not worth worrying about the number of steps in the computation (unless we are doing the calculation by hand). If we need the same sort of calculation many times then, though each step costs only a tiny amount of time and money, it may be important to seek an efficient method of computation.
(3) What can go wrong? If we are doing hand computations, it is possible to see problems developing and avoid them. Machines, although fast, are even stupider than we are and must be given instructions which cover every contingency.

In this chapter we discuss the problem of solving equations

$$f(x) = 0,$$

where f is well behaved. We shall see that even the simplest methods for solving this apparently simple problem require thought on our part.

The first problem that we face is that we (or our machines) only work to a certain degree of accuracy. When we compute $f(x)$, what we obtain is a number $F(x)$ with $|f(x) - F(x)| \leq u$, where u is some small positive number like 10^{-8} or 2^{-63}. This means that we can only guarantee that $f(x) > 0$ if $F(x) > u$ and that (unless we have other information) we can never say that $f(x) = 0$ but only that $|f(x)| \leq u$. For simplicity, we shall often ignore the limited accuracy and talk about $f(x)$ rather than $F(x)$, but the reader should always keep these accuracy limitations in mind.

The second problem that we need to consider is what happens if the equation has no solution in the range considered. If we instruct the computer to look for the non-existent solution, it will continue searching faithfully forever unless we have some appropriate stopping instruction. The third problem arises when there is more than one solution, in which case the solution delivered by the computer may not be the one we want.

The standard illustration of how these problems may arise and how they can interact is given by the case $f(x) = x^2 + v$ with v small. If $v > 0$ the equation $f(x) = 0$ has no roots. If $v < 0$ then the equation has two roots which are very close together. We should not be surprised if a computer (which can only work to a certain level of accuracy) will have great trouble trying to solve $x^2 = 0$.

Exercise 7.1.1. *Consider* $f(x) = (x - b)^3$. *If* $u > 0$, *find the* x *which satisfy* $|f(x)| \le u$. *Explain why this makes it hard to find* b *to a high degree of accuracy.*

Here is a simple method of root finding which depends on the observation that, if $a < b$ and $f(a) \le 0 \le f(b)$, there must exist a c with $a \le c \le b$ and $f(c) = 0$. For simplicity, we ignore the fact that our computations have limited accuracy. The method requires initial values a_0, b_0 such that $a_0 < b_0$ and $f(a_0) \le 0 \le f(b_0)$. Notice that this guarantees that there is a root between a_0 and b_0.

Let $c_0 = (a_0 + b_0)/2$. Either $f(c_0) \le 0$ and we set $a_1 = c_0$, $b_1 = b_0$ or $0 < f(c_0)$ and we set $a_1 = a_0$, $b_1 = c_0$. In either case, $f(a_1) \le 0 \le f(b_1)$ and $b_1 - a_1 = (b_0 - a_0)/2$ so we *have halved the length of the interval in which we know there exists a root*. We repeat the process as many times as we want obtaining intervals with end points a_j, b_j such that $f(a_j) \le 0 \le f(b_j)$ and $b_j - a_j = 2^{-j}(b_0 - a_0)$.

Exercise 7.1.2. *Write down explicitly how we obtain* a_{j+1} *and* b_{j+1} *from* a_j *and* b_j.

Exercise 7.1.3. *(i) Find* a_j, b_j *for* $j = 1, 2, 3, 4$ *when* $f(x) = x^2 - 4$ *and* $a_0 = 5/3$, $b_0 = 8/3$.
(ii) Find a_j, b_j *for* $j = 1, 2, 3, 4$ *when* $f(x) = x^2 - 2$ *and* $a_0 = 0$, $b_0 = 2$.

Exercise 7.1.4. *Explain why a similar method will work if we have* a_0, b_0 *such that* $a_0 < b_0$ *and* $f(b_0) \le 0 \le f(a_0)$.

The method just explained is called 'successive bisection' or, sometimes, 'lion hunting'.[1] If you only need to compute one root and you have a computer, it is a perfectly reasonable method.

Exercise 7.1.5. *Recall, or check, the very useful fact that* $2^{10} \approx 10^3$. *Deduce that, if* $b_0 = a_0 + 1$, *you can show that a root lies within a stated interval of length* 10^{-12} *by using* 40 *iterations (that is to say, halving the interval* 40 *times).*

7.2 The Newton–Raphson method

Successive bisection is tedious if you have to do it with pencil and paper. The Newton–Raphson method was developed in order to reduce the number of calculations required for finding the roots of polynomials and, more generally, finding a such that $f(a) = 0$ when f is a well understood function.

The idea is simple. Choose x_0 in the region where you guess the desired solution a will lie. Since f is well behaved,

$$f(x) \approx f(x_0) + f'(x_0)(x - x_0)$$

so we expect

$$0 = f(a) \approx f(x_0) + f'(x_0)(a - x_0)$$

whence, rearranging,

$$a \approx x_0 - \frac{f(x_0)}{f'(x_0)}.$$

(I think of this algebra as 'sliding down the tangent' as shown in Figure 7.1.) It therefore makes sense to choose our next guess to be

$$x_1 = x_0 - \frac{f(x_0)}{f'(x_0)}.$$

We then guess again, using x_1 as our initial guess, and then repeat the process as many times as desired. Thus, if our jth guess is x_j, our $j + 1$th guess is

$$x_{j+1} = x_j - \frac{f(x_j)}{f'(x_j)}.$$

This is the Newton–Raphson method.

Exercise 7.2.1. *(i) Find* x_j *for* $j = 1, 2, 3, 4$ *when* $f(x) = x^2 - 4$ *and* $x_0 = 3$. *(ii) Find* x_j *for* $j = 1, 2, 3, 4$ *when* $f(x) = x^2 - 2$ *and* $x_0 = 1$.

[1] This nomenclature may be traced back to Method 4 in the paper of H. Pétard entitled 'A contribution to the mathematical theory of big game hunting', *American Mathematical Monthly* 45 (1938) pp. 446–447.

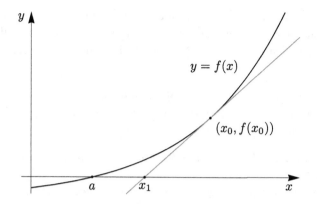

Figure 7.1 Sliding down the tangent

If we compare the results of Exercise 7.2.1 with those of Exercise 7.1.3 we see that, at least in the cases chosen, Newton–Raphson beats bisection hands down.

To see why this is so, we look at the Taylor series of the function

$$g(x) = x - \frac{f(x)}{f'(x)}.$$

We observe that

$$g'(x) = 1 - \frac{f'(x)}{f'(x)} + \frac{f(x)f''(x)}{f'(x)^2} = \frac{f(x)f''(x)}{f'(x)^2}$$

and so (since $f(a) = 0$)

$$g(a) = a, \ g'(a) = 0.$$

The local Taylor's theorem now tells us that there exist $u > 0$ and $M > 0$ such that

$$|g(x) - a| = |g(x) - g(a) - g'(a)(x - a)| \le M(x - a)^2$$

whenever $|x - a| \le u$. Thus, provided that $|x_j - a| \le u$,

$$|x_{j+1} - a| \le M|x_j - a|^2.$$

If the error $|x_j - a|$ is small, then the error $|x_{j+1} - a|$ is very small indeed.

Exercise 7.2.2. *People sometimes say that, if we use the Newton–Raphson method, the number of significant figures in the answer doubles at each step. Explain why this is more or less true in appropriate cases.*

Of course, the argument just given only tells us that appropriate u and M exist, but not how small they are. If our initial guess is too far from the root, there is no reason to expect the Newton–Raphson method to behave as we would wish.

Exercise 7.2.3. *Let us apply the Newton–Raphson method to $f(x) = \cos x$. Show that we can choose x_0 with $0 < x_0 < \pi/2$ such that $x_1 = x_0 + 2\pi$. With this choice, what is the value of x_j?*

Exercise 7.2.4. *Apply the Newton–Raphson method to $f(x) = x^3 - 2x + 2$ with $x_0 = 0$.*
[It is quite interesting to use a calculator and watch what happens when $x_0 = 10^{-2}$.]

Exercise 7.2.5. *(i) Suppose that $f(x) = x^{1/3}$. What happens if you apply the Newton–Raphson method with $x_0 \neq 0$? Why does our demonstration that the Newton–Raphson method works for a reasonable initial guess fail?*
(ii) Suppose now that F is a well behaved function everywhere, but that $F(x) = x^{1/3}$ for $|x| \geq u$. Show that the Newton–Raphson method for solving $F(x) = 0$ will fail if our starting point x_0 is such that $|x_0| > u$.

In practice, problems arising from a poor choice of x_0 can be detected by looking at the values of $|x_{j+1} - x_j|$ and halting the computation if these values do not decrease rapidly with j. A much more serious problem is that we cannot compute f' for many of the functions f that we meet. Thus the Newton–Raphson method can only be used in special (though important) cases such as when f is a polynomial.

Exercise 7.2.6. *The power series method gives a rapid way of computing $\exp x$ when $0 \leq x \leq 1$. Show how the Newton–Raphson method now gives us a fast method of computing $\log t$ for reasonable values of t.*

Exercise 7.2.7.•**[Halley's method]**[2] *(i) Let f be a well behaved function with $f(a) = 0$. We set*

$$g(x) = x - \frac{2f(x)f'(x)}{2f'(x)^2 - f(x)f''(x)}.$$

Take a large sheet of paper and calculate $g'(x)$. Show that $g'(a) = g''(a) = 0$. (Long calculations occur less frequently in mathematics than non-mathematicians believe, but they do occur.)

---

[2] Halley of Halley's comet, friend of Newton and all-round clever man.

(ii) Deduce that there exist u > 0 and K > 0 such that

$$|g(a+h) - a| \leq K|h|^3$$

whenever $|h| \leq u$.

(iii) Show that, if G(x) is the effect of applying the Newton–Raphson method twice to x (so, in our earlier notation, $G(x_0) = x_2$), then we can find $v > 0$ and $L > 0$ such that $|G(a+h) - a| \leq L|h|^4$ whenever $|h| \leq v$.

Thus, if we are sufficiently close to the root, applying the Newton–Raphson method twice will be more effective than applying the more complicated Halley's method once. This explains why Halley's method and its still more complicated higher order analogues are not much used.

7.3 There are lots of numbers

Most of my readers will be aware that the Ancient Greeks discovered that not all numbers are rational (that is to say, not all numbers can be written as p/q with p and q integers and $q > 0$). The first such number to be discovered was $\sqrt{2}$. Numbers that are not rational are called *irrational*.

Exercise 7.3.1. *(If you have not met the argument before or you need to be reminded of it.)*

 (i) *Suppose, if possible, that $\sqrt{2} = u/v$ with u and v integers having no common factor. Show that $u^2 = 2v^2$ and deduce that u is even, i.e. that $u = 2r$ for some integer r. Deduce that $v^2 = 2r^2$ and so that v is even and u and v have 2 as a common factor contradicting our original assumption.*
 (ii) *Extend the argument to show that, if p is prime, $\sqrt{p}$ is irrational.*
 (iii) *Extend the argument still further to show that, if a positive integer n is not the square of an integer, $\sqrt{n}$ is irrational.*

Many other irrational numbers can be produced in a similar way, but, until the middle of the nineteenth century, the only numbers which could be proved irrational were roots of polynomials of the form

$$b_n t^n + b_{n-1} t^{n-1} + \cdots + b_1 t + b_0$$

with $b_n \neq 0$ and all the b_j integers. Roots of such polynomials (whether rational or irrational) are called *algebraic* numbers.

Mathematicians suspected that numbers like π and $e = \exp 1$ were not algebraic, but could not show that *any* non-algebraic (or, as they are called, *transcendental*) numbers existed. In 1844, Liouville resolved the problem with a proof resembling one of those

> ... jewels five-words-long
> That on the stretched forefinger of all Time
> Sparkle forever ... [3]

In the previous part of this chapter, we used versions of the mean value inequality and Taylor's theorem to show that we can obtain good approximations to roots. Liouville turned this idea on its head to show that irrational algebraic numbers cannot be very well approximated by rational numbers. He then wrote down an irrational number which could be very well approximated by rationals and which must therefore be transcendental.

Here is his clever argument. For the moment, let us look at a particular polynomial

$$P(t) = b_n t^n + b_{n-1} t^{n-1} + \cdots + b_1 t + b_0$$

with $b_n \neq 0$ and all the b_j integers. Since a polynomial of degree n has at most n roots (see Exercise 7.3.6), we can find an R such that, if $P(x) = 0$, then $|x| \leq R - 2$. We can now find a $K \geq 1$ such that $|P'(t)| \leq K$ whenever $|t| \leq R$.

Exercise 7.3.2. *Show that the choice*

$$K = n|b_n|R^{n-1} + (n-1)|b_{n-1}|R^{n-2} + \cdots + 2|b_2|R + |b_1| + 1$$

will do.

The mean value inequality now gives

$$|P(t) - P(x)| \leq K|t - x|$$

for all t and x with $|t| \leq R$ and $|x| \leq R$.

Next we observe that, if p and q are integers with $q \geq 1$, then $q^n P(p/q)$ is an integer and, in particular, if $P(p/q) \neq 0$, we must have $q^n|P(p/q)| \geq 1$, so $|P(p/q)| \geq q^{-n}$. Combining this remark with the result of the previous paragraph, we see that, if x is a root of P, $P(p/q) \neq 0$ and $|p/q| \leq R$, we have

$$q^{-n} \leq |P(p/q)| = |P(p/q) - P(x)| \leq K|p/q - x|$$

---

[3] Tennyson, *The Princess.*

and so

$$|p/q - x| \geq K^{-1}q^{-n}.$$

We now do a bit of tidying up.

Exercise 7.3.3. *(i) (This is as trivial as it looks.) If x is a root of P, show that, if p and q are integers with q > 0 and $|p/q| > R$, then $|p/q - x| \geq K^{-1}q^{-n}$.*
(ii) Let x be an irrational root of P. Use the fact that there are only finitely many roots of P to show that there exists a q_0 such that $P(p/q) \neq 0$ whenever p and q are integers with $|p/q - x| < K^{-1}q^{-n}$ and $q \geq q_0$.

It is now easy to see that, if x is an irrational root of P, the inequality

$$|p/q - x| \leq q^{-n-1}$$

can have only finitely many solutions with p and q integers and $q \geq 1$. Since this statement does not depend on the P chosen, we have, in fact, proved that, if x is an irrational root of any polynomial of degree n with integer coefficients, the inequality

$$|p/q - x| \leq q^{-n-1}$$

can have only finitely many solutions with p and q integers and $q \geq 1$.

It now follows that, if x is an irrational number such that we can find integers p_n and $q_n \geq 2$ with

$$|p_n/q_n - x| \leq q_n^{-n-1}$$

for all $n \geq 1$, then x cannot be the root of any polynomial with integer coefficients, i.e. x must be transcendental.

We now write down such an x. Let x be the number with $0 < x < 1$ whose decimal expansion has 1 in the n!th place for each $n \geq 1$ and zero in every other place. Thus

$$x = .110\,001\,000\ldots.$$

Exercise 7.3.4. *Write out x to 30 places of decimals.*

If we now write $q_n = 10^{(n+2)!}$ and

$$p_n = 10^{(n+2)!-1!} + 10^{(n+2)!-2!} + 10^{(n+2)!-3!} + \cdots + 10^{(n+2)!-n!}$$
$$+ 10^{(n+2)!-(n+1)!} + 1,$$

we see, by looking at the decimal expansions of x and p_n/q_n, that

$$\frac{p_n}{q_n} \leq x \leq \frac{p_n}{q_n} + 2 \times 10^{-(n+3)!}$$

and so

$$\left| x - \frac{p_n}{q_n} \right| \leq 2 \times 10^{-(n+3)!} \leq q_n^{-n-1}$$

for every $n \geq 1$. Thus x is transcendental.

Exercise 7.3.5. *Let r_1, r_2, ...be integers with $0 \leq r_n \leq 9$ and let y be the number with $0 \leq y < 1$ whose decimal expansion has r_n in the $n!$th place for each $n \geq 1$ and zero in every other place. Explain why y is rational if all but finitely many of the r_j are zero. Show that, if this is not the case, then y is transcendental.*

We have shown that there are many transcendental numbers, but not that particular numbers like e and π are transcendental. Thirty years after Liouville's proof, Hermite showed that e is transcendental and later, building on Hermite's ideas, Lindemann showed that π is. Further work on transcendental numbers involved some of the deepest mathematics of the last century.

Exercise 7.3.6.• *If you are interested (but only if you are interested), this exercise gives a pretty demonstration that a polynomial of degree n can have at most n distinct roots.*

(i) *Let $n \geq 1$. Show that if P is a polynomial of degree n, that is to say*

$$P(t) = a_n t^n + a_{n-1} t^{n-1} + \cdots + a_0$$

with $a_n \neq 0$ and u is a given number, then we can find a polynomial

$$Q(t) = b_{n-1} t^{n-1} + b_{n-2} t^{n-2} + \cdots + b_0$$

of degree at most $n - 1$ and a number r such that

$$P(t) = (t - u)Q(t) + r.$$

(ii) *Continuing from (i), show that, if u is a root of P (that is to say, $P(u) = 0$), then $r = 0$. Deduce that*

$$P(t) = (t - u)Q(t)$$

and that, if v is a root of P, then either $v = u$ or v is a root of Q.

(iii) *Show that polynomial of degree n with $n \geq 1$ can have at most n distinct roots.*

(iv) *If P is a polynomial of degree at most n and P takes the value 0 at $n + 1$ distinct points, show that $P(t) = 0$ for all t. (We used this fact in the sentence following Exercise 3.1.2.)*

8

Hills and dales

8.1 More than one variable

Up to now, we have only looked at functions $f(x)$ of one variable x. However, this is inadequate for the study of the real world. If we look at an ideal gas, we have the equation

$$\frac{PV}{T} = R,$$

where P represents pressure, V represents volume, T represents temperature and R is a constant. Thus P depends on V and T, that is to say, P can be considered as a function of V and T. (Note that we could change our point of view and consider V as a function of P and T or T as a function of V and P.) If we want a more realistic model of a gas, we will have

$$P = f(V, T)$$

for some function f of two variables.

Other physical quantities will depend on position given by the three coordinates (x, y, z) and time t and so will be functions of *four* variables. The more variables we have, the more difficult it becomes to cope with the notation and the more difficult it becomes to visualise what is going on, so we shall stick with two variables and find this sufficiently hard.

In order to gain some insight, I shall picture functions of two variables by thinking of $f(x, y)$ as the height above sea level at the point with Cartesian coordinates (x, y).

In order to guide themselves round the landscape given by f, human beings use maps, that is to say two-dimensional representations of a three-dimensional reality. The earliest maps ignored hills altogether and later maps simply showed little pictures of hills. Such maps were useful for individual travellers but useless for military purposes, and in the next stage of map-making an, often very beautiful, system of hill-shading was used, copying the shadows that would be cast by an imaginary sun at a particular point in the sky. Inspection of such maps shows that they give a very clear idea of the steepness of slopes. The idea of drawing contours, that is to say curves along points at equal heights, seems to appear first in the work of the Dutch engineer Nicholas Cruquius, who used it to plot the depth of a river, and was gradually adopted over the next century.

If the reader thinks that the process of adoption was rather slow, she should note that it is not easy to explain precisely what a contour line is and that the interpretation of a contour map is a skill that has to be acquired.

Exercise 8.1.1. *Suppose that you are talking to an intelligent eighteenth-century general. Explain what a contour map is and how it is used. Why should he prefer it to maps using hill shading which he knows and understands? Why is the cost of surveying for a contour map likely to be higher?*

I shall make life easier for myself by assuming that the reader can read a contour map. (If not, she can certainly spend her time more profitably by learning to read a map than by reading this chapter.) I shall try to show that the calculus sheds light on certain features of the map.

What should it mean to say that a function of two variables is well behaved? Recall that we said that a function f of *one* variable was continuous if it was 'close to constant at all sufficiently fine scales' and we gave the following formal definition.

Definition 1.3.3. *A function f of one variable is continuous at t if, given any $u > 0$, we can find a $v > 0$ such that $|f(t + h) - f(t)| \le u$ whenever $|h| \le v$.*

There is no reason why we should not imitate our successful one-dimensional definition by a similar two-dimensional version. Thus we shall say that a function g of two variables is continuous if it is 'close to constant at all sufficiently fine scales' and give the following formal definition.

Definition 8.1.2. *A function g of two variables is continuous at (x, y) if, given any $u > 0$, we can find a $v > 0$ such that $|g(x + h, y + k) - g(x, y)| \le u$ whenever $(h^2 + k^2)^{1/2} \le v$.*

We have to think a bit more when it comes to differentiability. In one dimension we considered the approximate relation

$$f(t + h) \approx f(t) + Ah$$

which corresponds to saying that (if h is small) the graph of $y = f(t + h)$ looks like the line $y = f(t) + Ah$ as a function of h. It is natural to replace the line by a plane so that (if h and k are small) $g(x + h, y + k)$ looks like the plane $g(x, y) + Ah + Bk$ as a function of h and k. Thus we want

$$g(x + h, y + k) \approx g(x, y) + Ah + Bk.$$

As in the one-dimensional case, this is only going to give us useful information if we specify that $\approx$ should mean that the error

$$|g(x + h, y + k) - (g(x, y) + Ah + Bk)|$$

is small in comparison with $|h|$ and $|k|$.

Exercise 8.1.3. *Show that*

$$\max\{|h|, |k|\} \leq (h^2 + k^2)^{1/2} \leq 2^{1/2} \max\{|h|, |k|\}.$$

Looking at the exercise you have just done, we see that the preceding statement can be recast as saying that

$$\frac{|g(x + h, y + k) - (g(x, y) + Ah + Bk)|}{(h^2 + k^2)^{1/2}}$$

is small. More exactly, we mean that we can make

$$\frac{|g(x + h, y + k) - (g(x, y) + Ah + Bk)|}{(h^2 + k^2)^{1/2}}$$

as small as we like, provided that we take $(h^2 + k^2)^{1/2}$ sufficiently small. The next two paragraphs run in almost, but not quite exactly, the same way as in the one variable case.

If the conditions of the previous paragraph hold, we say that g is *differentiable*. Sometimes we shall write

$$g(x + h, y + k) = g(x, y) + Ah + Bk + o\big((h^2 + k^2)^{1/2}\big). \quad\bigstar$$

As the reader will expect, mathematicians pronounce '$+o\big((h^2 + k^2)^{1/2}\big)$' as 'plus little o of $(h^2 + k^2)^{1/2}$', but I very strongly recommend that she pronounces it as

'plus an error term which diminishes faster than linear'

or

'plus an error term which diminishes faster than $|h|$ and $|k|$'.

In advanced courses the notion of differentiability is made completely precise by using the following form of words.

Definition 8.1.4. *A function g of two variables is differentiable at* (x, y) *with derivative given by* (A, B) *if, given any* $u > 0$, *we can find a* $v > 0$ *such that*

$$\frac{|g(x + h, y + k) - (g(x, y) + Ah + Bk)|}{(h^2 + k^2)^{1/2}} \leq u$$

whenever $(h^2 + k^2)^{1/2} \leq v$.

In the informal language of Section 1.2,

$$g(x + h, y + k) = g(x, y) + Ah + Bk$$

to first order.

It turns out that A and B have an interpretation in terms of operations that we have already introduced. Suppose that we set $k = 0$ and allow h to vary. Then equation ★ becomes

$$g(x + h, y) = g(x, y) + Ah + o(h),$$

so the function f_y of one variable, defined by $f_y(x) = g(x, y)$, is differentiable with $f'_y(x) = A$. We shall write $\partial_1 g(x, y) = f'_y(x)$. The *partial derivative* $\partial_1 g$ (pronounced 'partial d one of g' or 'curly d one of g') is thus obtained by keeping the second variable fixed and differentiating with respect to the first variable.[1]

Exercise 8.1.5. *Define* $\partial_2 g$ *in a similar way and show that*

$$g(x + h, y + k) = g(x, y) + \big(\partial_1 g(x, y)h + \partial_2 g(x, y)k\big) + o\big((h^2 + k^2)^{1/2}\big).$$

8.2 Taylor's theorem in two variables

Earlier, when we studied functions of one variable, we saw that, once we had defined the notion of a differentiable function (that is to say, a locally almost linear function), we could use repeated differentiation to obtain information about the quadratic, cubic and higher order local behaviour of smooth functions via the local version of Taylor's theorem. Can we obtain similar results in higher dimensions (that is to say, with more variables)?

One way to do so is to reduce our problem to a one-dimensional one. Suppose that g is a well behaved function of two variables. If we define

$$f_\theta(r) = g(r \cos \theta, r \sin \theta),$$

---

[1] There are several different notations for partial derivatives. I have chosen one of the less usual because it requires less preliminary discussion.

then $f_\theta(r)$ is a well behaved function of the single variable r and we can apply the local version of Taylor's theorem (see ★ on page 100) for one variable to get

$$f_\theta(r) = f_\theta(0) + \frac{f'_\theta(0)}{1!}r + \frac{f''_\theta(0)}{2!}r^2 + \cdots + \frac{f_\theta^{(n-1)}(0)}{(n-1)!}r^{n-1} + o(r^{n-1}).$$

To get a clearer picture, think of a walker lost in a fog. The walker places a large white stone at a fixed point $(0, 0)$ and then walks in a straight line in a particular direction to the point $(r\cos\theta, r\sin\theta)$. She then returns to her starting point and sets out in another direction. By repeating this process several times, she can get a good idea of what the hillside looks like close to her.

Of course, the formula just obtained is no use unless we can find the derivatives of f in terms of the partial derivatives of g. We observe[2] that

$$\begin{aligned}
f_\theta(r + h) &= g\big((r + h)\cos\theta, (r + h)\sin\theta\big)\\
&= g(r\cos\theta + h\cos\theta, r\sin\theta + h\sin\theta)\\
&= g(r\cos\theta, r\sin\theta)\\
&\quad + \partial_1 g(r\cos\theta, r\sin\theta)h\cos\theta + \partial_2 g(r\cos\theta, r\sin\theta)h\sin\theta\\
&\quad + o\big((h^2(\cos\theta)^2 + h^2(\sin\theta)^2)^{1/2}\big)\\
&= f_\theta(r) + \big(\partial_1 g(r\cos\theta, r\sin\theta)\cos\theta\\
&\quad + \partial_2 g(r\cos\theta, r\sin\theta)\sin\theta\big)h + o(h)
\end{aligned}$$

and so

$$f'_\theta(r) = \partial_1 g(r\cos\theta, r\sin\theta)\cos\theta + \partial_2 g(r\cos\theta, r\sin\theta)\sin\theta.$$

To get the second derivative we write

$$G_1(x, y) = \partial_1 g(x, y), \quad G_2(x, y) = \partial_2 g(x, y)$$

and

$$G_{1,\theta}(r) = G_1(r\cos\theta, r\sin\theta), \quad G_{2,\theta}(r) = G_2(r\cos\theta, r\sin\theta).$$

Then

$$f'_\theta(r) = G_{1,\theta}(r)\cos\theta + G_{2,\theta}(r)\sin\theta$$

---

[2] The reader may recoil in horror from the formulae that follow, but they are merely simple observations dressed up in scary notation.

and so

$$f_\theta''(r) = G_{1,\theta}'(r)\cos\theta + G_{2,\theta}'(r)\sin\theta$$

$$= \big(\partial_1 G_1(r\cos\theta, r\sin\theta)\cos\theta + \partial_2 G_1(r\cos\theta, r\sin\theta)\sin\theta\big)\cos\theta$$

$$+ \big(\partial_1 G_2(r\cos\theta, r\sin\theta)\cos\theta + \partial_2 G_2(r\cos\theta, r\sin\theta)\cos\theta\big)\sin\theta$$

$$= \partial_1\partial_1 g(r\cos\theta, r\sin\theta)(\cos\theta)^2 + \partial_2\partial_1 g(r\cos\theta, r\sin\theta)\sin\theta\cos\theta$$

$$+ \partial_1\partial_2 g(r\cos\theta, r\sin\theta)\cos\theta\sin\theta + \partial_2\partial_2 g(r\cos\theta, r\sin\theta)(\sin\theta)^2.$$

Since we are only interested in quadratic approximation, we stop here. Choosing θ and r so that $h = r\cos\theta$ and $k = r\sin\theta$, we obtain

$$g(h,k) = g(r\cos\theta, r\sin\theta) = f_\theta(r)$$

$$= f_\theta(0) + f_\theta'(0)r + \frac{f_\theta''(0)}{2}r^2 + o(r^2)$$

$$= g(0,0) + (\partial_1 g(0,0)\cos\theta + \partial_2 g(0,0)\sin\theta)r$$

$$+ \frac{1}{2}\bigg(\partial_1\partial_1 g(0,0)(\cos\theta)^2 + \partial_2\partial_1 g(0,0)\sin\theta\cos\theta$$

$$+ \partial_1\partial_2 g(0,0)\cos\theta\sin\theta + \partial_2\partial_2 g(0,0)(\sin\theta)^2\bigg)r^2 + o(r^2)$$

$$= g(0,0) + \partial_1 g(0,0)h + \partial_2 g(0,0)k$$

$$+ \frac{1}{2}\bigg(\partial_1\partial_1 g(0,0)h^2 + \big(\partial_2\partial_1 g(0,0) + \partial_1\partial_2 g(0,0)\big)hk$$

$$+ \partial_2\partial_2 g(0,0)k^2\bigg) + o(h^2 + k^2).$$

Exercise 8.2.1. *Write down a similar formula for $g(x + h, y + k)$.*

Exercise 8.2.2. *Suppose that*

$$G(x, y) = a_0 + b_1 x + b_2 y + c_{11}x^2 + 2c_{12}xy + c_{22}y^2.$$

Compute

$$\partial_1 G(x,y),\ \partial_2 G(x,y),\ \partial_1\partial_1 G(x,y),\ \partial_1\partial_2 G(x,y),\ \partial_2\partial_1 G(x,y),\ \partial_2\partial_2 G(x,y).$$

Verify that

$$G(h,k) = G(0,0) + \partial_1 G(0,0)h + \partial_2 G(0,0)k$$

$$+ \frac{1}{2}\bigg(\partial_1\partial_1 G(0,0)h^2 + \big(\partial_2\partial_1 G(0,0)$$

$$+ \partial_1\partial_2 G(0,0)\big)hk + \partial_2\partial_2 G(0,0)k^2\bigg).$$

Remark. It is a remarkable fact that $\partial_1\partial_2 G = \partial_2\partial_1 G$ for well behaved functions G. However, the arguments of this chapter do not give any reason for believing this and we shall make no use of the fact.[3]

Exercise 8.2.3.● *We have obtained the* quadratic *local Taylor's expansion of $g(h, k)$. We shall not need it, but it is an 'instructive exercise' to obtain the* cubic *local Taylor's expansion. (You will need a large sheet of paper or better notation.) Verify your answer in the case of a cubic multinomial in the same way as you verified the quadratic expansion in Exercise 8.2.2.*

The reader may also ask, particularly if she has done Exercise 8.2.3, if we could not find a better abbreviated notation for Taylor series. The answer is that we could, but that such a notation would require time to set up and understand.

For the purposes of this chapter, we concentrate on the fact that, for a well behaved function g, we have

$$g(h, k) = C + (Ak + Bh) + \frac{1}{2}(ah^2 + 2bhk + ch^2) + o(h^2 + k^2)$$

for appropriate constants $C = g(0, 0)$, $A = \partial_1 G(0, 0)$, $B = \partial_2 G(0, 0)$, a, b and c. If we replace $g(h, k)$ by $g(h, k) - C$ (that is to say, readjust the zero point from which we measure height) we can reduce the number of constants and suppose that

$$g(h, k) = (Ak + Bh) + \frac{1}{2}(ah^2 + 2bhk + ch^2) + o(h^2 + k^2).$$

In general we can expect that at least one of A and B is non-zero. In this case, we observe that, automatically,

$$g(h, k) = Ak + Bh + o\big((h^2 + k^2)^{1/2}\big),$$

and so the contour lines $g(h, k) = q$ look very much like

$$Ak + Bh = q.$$

Thus, in general (specifically, in the cases when at least one of $\partial_1 g(0, 0)$ and $\partial_2 g(0, 0)$ is non-zero), we may expect that, at a sufficiently fine scale, the contour lines will look like equally spaced parallel lines.

What about the special case when $\partial_1 g(0, 0) = \partial_2 g(0, 0) = 0$, so that $A = B = 0$ and

$$g(h, k) = \frac{1}{2}(ah^2 + 2bhk + ch^2) + o(h^2 + k^2)?$$

---

[3] It turns out that the proof involves sending our fog-bound walker along two different routes to the same point.

(If this is the case, we say that $(0, 0)$ is a *critical point*.) The expression just stated can be simplified by rotating the system.

Exercise 8.2.4. *If we rotate the plane through an angle θ keeping the origin fixed, show that the point (x, y) is carried to the point*

$$(x \cos \theta - y \sin \theta, x \sin \theta + y \cos \theta).$$

Let us write

$$G(s, t) = g(s \cos \theta - t \sin \theta, s \sin \theta + t \cos \theta).$$

Then

$$G(s, t) = \frac{1}{2}\big(a(s \cos \theta - t \sin \theta)^2 + 2b(s \cos \theta - t \sin \theta)(s \sin \theta + t \cos \theta)$$
$$+ c(s \sin \theta + t \cos \theta)^2\big) + o(s^2 + t^2)$$
$$= \frac{1}{2}(ut^2 + 2vst + ws^2) + o(s^2 + t^2),$$

where (using the double angle formulae of Exercise 1.5.3 (iv))

$$2v = 2(c - a)\cos \theta \sin \theta + 2b\big((\cos \theta)^2 - (\sin \theta)^2\big)$$
$$= (c - a)\sin 2\theta + 2b \cos 2\theta.$$

By choosing θ appropriately we can make $v = 0$.

Exercise 8.2.5. *(i) Check the formula for v. Explain why you can always find a θ such that $v = 0$.*

(ii) We do not need to find expressions for u and w, but you may wish to do so as an exercise.

By rotating our map, we have reduced our study of contour lines near a critical point to the case when

$$g(h, k) = \frac{1}{2}(Ph^2 + Qk^2) + o(h^2 + k^2)$$

for appropriate P and Q. We shall not look at the very special cases when either of P and Q vanishes, but assume that P, $Q \neq 0$. We see that there are three possibilities.

Case 1. If P, $Q < 0$, then g attains a (local) maximum at $(0, 0)$. The contour lines are approximate ellipses getting closer together as we move away from $(0, 0)$.

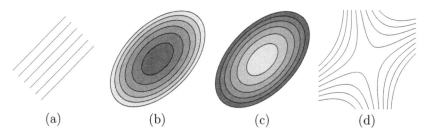

Figure 8.1 (a) Non-critical point, (b) maximum, (c) minimum, (d) saddle

Case 2. If P, $Q > 0$, then g attains a (local) minimum at $(0, 0)$. The contour lines are approximate ellipses getting closer together as we move away from $(0, 0)$.

Case 3. In the remaining case P and Q have opposite signs. We may suppose that $P > 0 > Q$. Then anyone travelling along the straight line path $k = 0$ will be at their lowest point when $h = 0$ and anyone travelling along to straight line path $h = 0$ will be at their highest point when $k = 0$. The contour lines of the form $g(x, y) = q$ with $q > 0$ are approximate hyperbolas getting closer together as we move away from $(0, 0)$ and the same is true for contour lines of the form $g(x, y) = q$ with $q < 0$.

Exercise 8.2.6. *Sketch the hyperbolas $x^2 - y^2 = 1$ and $x^2 - y^2 = -1$.*

Exercise 8.2.7. *The arguments and language of this chapter are fairly informal. Suppose that $f(x, y)$ is a function of two variables defined for $x^2 + y^2 \leq 1$ (and possibly elsewhere). Produce a mathematical description along the lines of the paragraph following Exercise 2.5.4 of what it means to say that f has a local maximum at $(0, 0)$.*

We have shown that (ignoring the unusual cases when P or Q vanishes) there are three types of critical points. Case 1 corresponds to the summit of a hill, Case 2 to the bottom of a depression and Case 3 to a pass (often called a *saddle point* since the surrounding surface looks a bit like a saddle).

We sketch the various possible contours in Figure 8.1.

Our discussion shows that the tops of hills, the bottoms of depressions and passes will all be pretty flat. The reader may object that the summits of mountains like Mount Everest are not flat, but we are talking about smooth functions, so our discussion is relevant to rolling hills rather than jagged mountains. Hill walkers are well aware that (at least in a rainy country like England) the tops of hills and passes (saddle points) tend to be marshy since the absence of a significant slope means that they are badly drained.

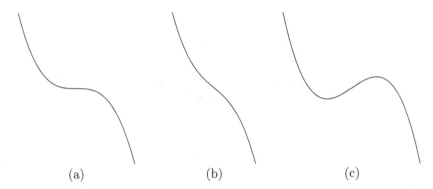

Figure 8.2 (a) Stationary point of inflexion, (b) no stationary point, (c) maximum and minimum

Exercise 8.2.8. *The length of the day (more precisely, the length of daylight) varies throughout the year. Explain, using simple calculus ideas, why the change in the length of days is small at dates close to the longest and shortest days and larger at other times.*

8.3 On the persistence of passes

In this section I shall use particularly informal arguments.[4] My object is not to convince readers that certain things are true, but to show that something along similar lines ought to be true.

When we looked at the one-dimensional case in Section 2.5 and again on page 100, we saw that, usually, when $f'(0) = 0$ the derivative $f'(t)$ changed sign as t went from negative to positive. and we obtained either a maximum or a minimum. However, it was possible for the sign to remain unchanged giving a so called 'stationary point of inflexion'. Such points are, in some vague sense, accidental because they are destroyed by small changes in f. Figure 8.2 shows that (again in some vague sense) we might expect small changes either to leave no stationary[5] points or to produce one maximum and one minimum.

Things are very different if we work in two dimensions.

It is geographically obvious that small changes in the landscape may make small changes in the positions and heights of peaks, passes, and low points,[6]

---

[4] I have been told that Grothendieck coined the term 'Mathematics fiction' by analogy with 'Science fiction' for mathematical speculations which, without being exact enough to prove or disprove, are worth thinking about.

[5] The reader may ask why we do not talk about 'critical' rather than 'stationary' points? She is right, but the usage is traditional.

[6] There are many words in English for high points and rather few for low points, doubtless because most low points are under water.

Figure 8.3 Two lakes coalesce

Figure 8.4 An island forms

but will not affect their nature. Here is a result discovered by Maxwell which shows that passes are an essential part of the two-variable landscape.

In view of Grothendieck's phrase, there is a happy science fiction air to our argument. Suppose that we drain away all the water from the earth. There will be V peaks, E passes and F low points. We do a certain amount of earth moving so as to raise peaks, lower low points and modify passes so that all the peaks have the same height, all the low points have the same height (above the centre of the earth), all the passes have *different* heights, the highest pass is lower than the height of the summits and the lowest pass is higher than the height of the low points.

We now arrange for it to rain in such a way that the water level is the same everywhere. After a little rain we will have F lakes in one connected piece of dry land (which we call an island). As the rain continues, some of the passes will disappear under the water and we will see a Finnish landscape with lakes containing islands which contain lakes which contain islands and so on. When the last pass is gone we will have one lake and V waterless islands.

When a pass disappears, one of two things happens. In the first case, two *different* lakes will coalesce as in Figure 8.3 to form one. Both the number of passes and the number of lakes has been diminished by 1, but the number of islands remains the same.

In the second case, two arms of the *same* lake will join as in Figure 8.4 to give a new island. The number of passes has been diminished by 1 and the number of islands has been increased by 1, but the number of lakes remains the same.

In both cases,

number of passes + number of islands − number of lakes

remains the same. But we know that, initially,

number of passes + number of islands − number of lakes $= E + 1 - F$

and, at the end,

number of passes + number of islands − number of lakes
$$= 0 + V - 1 = V - 1.$$

Thus

$$E + 1 - F = V - 1$$

or, rearranging, $V + F - E = 2$. If we have V peaks and F low points we must have $V + F - 2$ passes. Saddle points are a necessary feature of two variable calculus.

Exercise 8.3.1. *What is the relation between the number of peaks, passes and low points in a continent like Australia?*

If the reader is prepared to accept a *very* informal discussion, we can recover Euler's formula for a polyhedron.[7] Observe that a tetrahedron has four vertices, six edges and four faces and a cube has eight vertices, six edges and six faces.

Exercise 8.3.2. *Check the statements just made and do similar counts for a few other polyhedrons.*

Euler's formula relates the number of vertices and edges to the number of faces. Distort your polyhedron so that each vertex becomes a peak, the centre of each face is a low point, the centre of each edge marks a saddle and these are the only peaks, low points and saddles. Our formula $V + F - E = 2$ now gives

number of vertices + number of faces − number of edges $= 2$.

Before leaving the subject of hills and dales, let us return to our walker wandering about in a fog. It is very easy for her to find a peak (that is to say, a local maximum) by using the obvious recipe:- 'Head upwards in the direction of greatest slope. If the slope is zero in all directions, put a large stone down as a marker and walk in a circle with centre the stone. If the ground slopes

---

[7] *Technical note.* The modern definition of a polyhedron is rather general, so we should really talk about a *convex* polyhedron.

downwards at every point on your path, you are at the top of a hill. If not, you should head upwards along the direction of greatest slope that you meet in your path.'

It is fairly clear that this recipe will get the walker to some peak (that is to say, local maximum), but that the only reason she might have for hoping that it is the highest peak (the global maximum) is the feeble one that the global maximum is also a local maximum. Of course, if she had a helicopter and there was no fog she could survey the entire countryside and so find the highest peak.[8]

In most cases that arise in practice, there is plenty of fog and no helicopter. If we are trying to find global maxima with a computer, we may adopt a strategy of distributing walkers over the countryside, letting each walker find a peak and hoping that the highest peak thus found is indeed the highest peak in the countryside.

If we seek to understand nature, the distinction may not matter since the systems studied may follow the same recipe as our walker and simply find a local maximum. The system that makes up a house seeks to minimise potential energy. The global minimum for such a system is a pile of rubble, but the local minimum at which it finds itself is attained by being a house. In evolution, being a shark is the best solution for being shark-like. Small changes in sharkiness reduce fitness, so the shark species remains unchanged. If a species of apes lives in trees and the trees gradually die then the old local maximum disappears, some small changes away from 'tree-living ape' may be advantageous and the species may move up a new slope to a new local maximum. It is possible to persuade oneself that similar considerations apply in economics and politics, but the analogy is so seductive that one must be careful not to overuse it.

[8] If we had some ham we could have some ham and eggs, if we had some eggs.

9

Differential equations via computers

9.1 Firing tables

When we looked at the paths of projectiles, we were able to solve the differential equations for the case when air resistance was negligible. It was clear that, if we added extra complications, there might be no explicit formula describing the path.

Until the end of the nineteenth century, artillery was directed by eye. The new guns of the beginning of the twentieth century had longer ranges and this meant that gunners fired at targets given by map references. Observers closer to the target might be able to observe the results and allow the gunners to adjust their aim. However, every time a gun fires it reveals both its position and its likely target. To avoid this, guns were issued with *firing tables* showing where shells were calculated to land. Just as in Exercise 4.3.4, this requires finding the solution of a pair of differential equations

$$x''(t) = -kx'(t), \quad y''(t) = -ky'(t) - g,$$

but now k is *not constant*, but itself a function of $v(t) = \sqrt{x'(t)^2 + y'(t)^2}$, the velocity of the shell and $y(t)$ its height (because of changes in the density of air). This function k also changes depending on other factors including the weather and the characteristics of the shell used. Besides the angle of elevation of the gun, the initial conditions (that is to say, $x'(0)$ and $y'(0)$) also depend on the temperature of the propellant used (which changes how vigorously it explodes), the number of previous firings (because of wear on the barrel) and so on.

Since there is no simple formula to describe the path, it is necessary to use a step-by-step approximate solution of the type we shall describe later in this chapter. It took a human being armed with a mechanical calculator an average of two eight-hour days to compute a single trajectory and each type of gun required the computation of hundreds of such trajectories.[1] It is no accident that the first general purpose electronic computer built in the USA was used for computing firing tables. The ENIAC could calculate the path of a shell in less time than the shell took in flight.

9.2 Euler's method

How do we (or, rather, our computers) find an approximate solution to a differential equation? We discuss the solution of the differential equation

$$y'(t) = f(y(t), t),$$

where f is a well behaved function. We shall use the 'initial condition' $y(0) = y_0$.

Exercise 9.2.1. *Check that the following differential equations have the form suggested:–*

$$y'(t) = ty(t), \ y'(t)\big(1 + y(t)^2\big) = 1, \ y'(t) = t^3 \sin(y(t)).$$

The most direct attack was suggested by Euler. We choose a small $h > 0$ (the step length) and use the approximation

$$y'(t) \approx \frac{y(t + h) - y(t)}{h}$$

to obtain

$$y(t + h) - y(t) \approx hf(y(t), t).$$

This suggests that we solve the exact equations

$$y_{r+1} - y_r = hf(y_r, rh)$$

and *hope* that

$$y(rh) \approx y_r.$$

Exercise 9.2.2. *It is a good idea to try writing a program to implement Euler's method on a computer.[2] However, we can also learn a lot by looking at problems which we can solve by algebra. (Both parts of this question provide useful information for the discussion that follows.) We take $T > 0$ and $h > 0$.*

---

[1] These figures are taken from 'Before the ENIAC' by H. Polachek in *IEEE Annals of the History of Computing* Volume 19, 1997.

[2] The author of this book is so old that the computing part of his undergraduate course was examined by asking 'How would you calculate X if you had a computer?'

(i) If $f(y, t) = t$ and $y_0 = 0$, show that $y_n = \frac{1}{2}n(n-1)h^2$ (see Exercise 2.2.11 (i) if necessary). In particular, if N is a strictly positive integer and $h = T/N$, show that

$$|y_N - y(T)| = Th/2.$$

(ii) If $f(y, t) = by$, show that $y_n = y_0(1 + bh)^n$. Deduce that, if T is fixed, N is a large positive integer and $h = T/N$, then

$$y_N \approx y_0 e^{bT}.$$

Why is this reassuring? (We shall use the approximation found in Section 5.1 repeatedly throughout this chapter.)

Banach used to say 'Hope is the mother of fools.' Why should we expect Euler's method to work? Assuming that the true solution y is such that $|y''(t)| \le M_1$ for all t in the range that we are interested in, Taylor's theorem with error estimate gives

$$|y(s + h) - y(s) - hf(y(s), s)| = |y(s + h) - y(s) - hy'(s)| \le M_1 \frac{h^2}{2}, \quad \bigstar$$

so making h small certainly improves 'local accuracy', but we should remember that the smaller h is, the more steps we have to take. Even more importantly, as y_s drifts away from $y(sh)$, the actual relevance of the inequality $\bigstar$ becomes less and less clear.

The only way forward is to roll up our sleeves and get down to some hard work. If the reader just reads the arguments[3] to gain an impression of what is going on, she will not lose much.

Let us write $e_r = y(rh) - y_r$ so that e_r is the error in our estimate of the true solution y at 'time' $t = rh$. Applying the standard trick of writing $0 = 1 - 1$, we have

$$
\begin{aligned}
|e_{r+1}| &= |y((r+1)h) - y_{r+1}| = |e_r + (y((r+1)h) - y(rh)) + (y_r - y_{r+1})| \\
&= |e_r + (y((r+1)h) - y(rh)) - hf(y_r, rh)| \\
&= |e_r + (y((r+1)h) - y(rh) - hf(y(rh), h)) \\
&\quad + h(f(y(rh), h) - f(y_r, rh))| \\
&\le |e_r| + |y((r+1)h) - y(rh) - hf(y(rh), h)| \\
&\quad + |h||f(y(rh), rh) - f(y_r, rh)|.
\end{aligned}
$$

---

[3] 'I like work: it fascinates me. I can sit and look at it for hours.' Jerome K. Jerome, *Three Men in a Boat*.

Before the reader's eyes glaze over, she should observe that the inequality says that the magnitude of error at $t = (r + 1)h$ is less than the sum of the three terms in the final line. The first term is the magnitude of the error at $t = rh$. The second term bounds the 'local error' due to replacing the differential equation by an approximate equation. We can estimate this term by using the inequality ★. The third term is the error incurred by the fact that we have wandered off course and, instead of starting the step at $y = y(rh)$, we start it at $y = y_r$. It is this term which concerns us most.

We need the idea of partial differentiation from Chapter 8 together with the mean value inequality. Suppose that $|\partial_1 f(y, s)| \leq M_2$ over the region that we are interested in. The mean value inequality gives

$$|f(u, s) - f(v, s)| \leq M_2|u - v|,$$

so

$$|f(y(rh), rh) - f(y_r, rh)| \leq M_2|y(rh) - y_r| = M_2|e_r|.$$

We now have estimates for the second and third terms in our inequality giving

$$|e_{r+1}| \leq |e_r| + M_1\frac{h^2}{2} + M_2h|e_r| = (1 + M_2h)|e_r| + M_1\frac{h^2}{2}.$$

We now apply the same kind of idea as we used to solve equation ★ on page 77. Instead of looking at $|e_r|$, we look at $w_r = (1 + M_2h)^{-r}|e_r|$, so that the inequality which concluded the previous paragraph becomes

$$w_{r+1} \leq w_r + v(1 + M_2h)^{-r},$$

where

$$v = \frac{M_1h^2}{2(1 + M_2h)}.$$

We then have

$$w_N \leq w_{N-1} + v(1 + M_2h)^{-(N-1)}$$
$$\leq w_{N-2} + v(1 + M_2h)^{-(N-2)} + v(1 + M_2h)^{-(N-1)}$$

$$\vdots$$

$$\leq w_0 + v\left(1 + (1 + M_2h)^{-1} + (1 + M_2h)^{-2} + \cdots + (1 + M_2h)^{-(N-1)}\right)$$
$$= 0 + v\frac{1 - (1 + M_2h)^{-N}}{1 - (1 + M_2h)^{-1}} = v\frac{1 - (1 + M_2h)^{-N}}{1 - (1 + M_2h)^{-1}}$$
$$= v(1 + M_2h)\frac{1 - (1 + M_2h)^{-N}}{M_2h}.$$

If we multiply through by $(1 + M_2h)^N$ and recall the definitions of w_n and v, we get

$$|e_N| \le (1 + M_2h)v \frac{(1 + M_2h)^N - 1}{M_2h} = \frac{M_1h}{2M_2}((1 + M_2h)^N - 1).$$

Now suppose we want to estimate the value of $y(T)$ by applying Euler's method with $h = T/N$. If N is large, then

$$(1 + M_2h)^N = \left(1 + \frac{M_2T}{N}\right)^N \approx \exp(M_2T)$$

and so the error $y(a) - y_N = e_N$ cannot exceed roughly

$$\frac{M_1}{2M_2}(\exp(M_2T) - 1)h,$$

that is to say, some constant multiple of h.

We have shown that, provided we choose the step length h small enough, the Euler method gives an estimate for $y(T)$ with an error which decreases at least as fast as (some constant multiple) of h. Since Exercise 9.2.2 (i) shows that in some cases the error does, indeed, decrease like a constant multiple of h, we cannot hope to do better.

Exercise 9.2.3. *Apply Euler's method with $h = 1/4$ to the equation*

$$y'(t) = -8y(t)$$

with initial value $y(0) = 1$. Why does the observed result not contradict our conclusions about Euler's method?

Exercise 9.2.4. Either *Write a program to implement Euler's method and apply it to the differential equation*

$$y'(t) = \frac{1}{3}y(t)$$

with $y(0) = 1$, $h = 1/10$ and $T = 10$. In order to produce a comparison with the later Exercise 9.3.2, you should a tabulate y_r and $y(rh)$ for step size $h = 1/10$ and $0 \le r \le 100$. Experiment with the effect of changing h whilst keeping T fixed.

Or *Use a calculator to work out the result of using Euler's method with $h = 1/10$ and $T = 2$, computing y_r for $0 \le y \le 20$ and comparing this with the true result $y(rh)$.*

9.3 A good idea badly implemented

It is very reassuring to know that we can solve differential equations to any required degree of accuracy by a step-by-step method on a computer. However, when we look at the details of the result obtained in the last section, we may be less happy. We showed that the errors in our approximation decreased at the same rate as the step length, but this means that to increase accuracy by a factor of 100 we must decrease step length by a factor of 100 and this means that we must make 100 times as many calculations which will take 100 times as long and cost 100 times as much.[4] As a US senator said in a similar case 'A billion dollars here, a billion dollars there and pretty soon you are talking real money.'[5]

Can we do better? Recall that, for Euler's method, the 'local accuracy', as estimated in ★, was bounded by a multiple of h^2, where h is the step length. If we could replace h^2 by some higher power of h, then we could hope for a corresponding increase in overall accuracy.

If we think about Taylor's theorem, it is not hard to come up with schemes which have higher local accuracy. One such scheme is the direct mid-point method.[6] We choose a small $h > 0$ (the step length) and, in place of the previous unsymmetrical approximation $y'(t) = (y(t + h) - y(t))/h$, use the symmetric approximation

$$y'(t) \approx \frac{y(t + h) - y(t - h)}{2h}$$

to obtain

$$y(t + h) - y(t - h) \approx 2hf(y(t), t).$$

We now solve the exact equations

$$y_{r+1} - y_{r-1} = 2hf(y_r, rh)$$

and *hope* that

$$y(rh) \approx y_r.$$

---

[4] Moreover, although we have not discussed this, the fact that the machine only calculates to a certain number of significant figures means that, if you make too many calculations, you begin to lose rather than gain accuracy.

[5] Senator Dirksen may have been misquoted, but found the phrase so apt that he never bothered to correct it.

[6] This is non-standard nomenclature, but there seems to be no agreement on an appropriate name for this method.

Assuming that the true solution y is such that $|y'''(t)| \leq M$ for all t in the range that we are interested in, Taylor's theorem with error estimate gives

$$\left| y(s+h) - y(s) - y'(s)h - \frac{y''(s)}{2}h^2 \right| \leq \frac{Mh^3}{6},$$

$$\left| y(s-h) - y(s) + y'(s)h - \frac{y''(s)}{2}h^2 \right| \leq \frac{Mh^3}{6}.$$

Thus

$$|y(s+h) - y(s-h) - 2hy'(s)|$$

$$= \left| \left(y(s+h) - y(s) - y'(s)h - \frac{y''(s)}{2}h^2 \right) \right.$$

$$\left. - \left(y(s-h) - y(s) + y'(s)h - \frac{y''(s)}{2}h^2 \right) \right|$$

$$\leq \left| y(s+h) - y(s) - y'(s)h - \frac{y''(s)}{2}h^2 \right|$$

$$+ \left| y(s-h) - y(s) + y'(s)h - \frac{y''(s)}{2}h^2 \right| \leq \frac{Mh^3}{3}$$

and the local error decreases with h^3.

Exercise 9.3.1. *Suppose that* $|y^{(5)}(t)| \leq M$ *over the range of interest. Find A and B such that*

$$|Ay(s+2h) + By(s+h) - hy'(s) - By(s-h) - Ay(s-2h)| \leq KMh^5$$

for some appropriate constant K. (As elsewhere in this chapter we take $h > 0$. *If we do not, we must replace* h^5 *by* $|h|^5$.)

The reader may object that the direct mid-point method requires us to solve

$$y_{r+2} - y_r = 2hf(y_{r+1}, (r+1)h)$$

and that, in order to start, we need to know, not only y_0, which we are given, but y_1, which we are not. However, we could calculate y_1 using Euler's method or otherwise. (Since we are only doing this calculation *once* we can use methods which require more work.) It is really not a problem and we shall assume an accurate y_1 from now on.

Exercise 9.3.2. Either *Write a program to implement the direct mid-point method and apply it to the differential equation*

$$y'(t) = \frac{1}{3}y(t)$$

with $y(0) = 1$. *(Just set* $y_1 = \exp h/3$.)

Tabulate the result y_r for step size $h = 1/10$ and $0 \le r \le 100$ and check that you do indeed get a better result than with Euler's method (see Exercise 9.2.4). Experiment with the effect of changing h.

Or *Use a calculator to work out the result of using the direct mid-point method method for step size $h = 1/10$ for $0 \le r \le 20$ and compare with the true result and with the result obtained using Euler's method.*

Or *Accept my word that the direct mid-point method gives a better result than the Euler method in this case.*[7]

So far, everything has worked out as it was supposed to. The first sign of trouble occurs when we seek to solve equations like $y'(t) = -3y(t)$.

Exercise 9.3.3. *Repeat Exercises 9.2.4 and 9.3.2 with the differential equation*

$$y'(t) = -3y(t)$$

and $y(0) = 1$. (Just set $y_1 = \exp(-3h)$ for the direct mid-point method.) Tabulate $\exp(-3rh)$, the result using Euler's method and the result of using the direct mid-point method for step size $h = 1/10$ and $0 \le r \le 20$ if you are using a calculator and $0 \le r \le 100$ if you are using a computer.

If you are using a computer, experiment with smaller values of h keeping T fixed. (If your machine is very accurate, you may wish to take T somewhat larger than 10.)

At first, the direct mid-point solution tracks the true solution rather well, but, at some point, it starts to diverge from the true solution and then, quite soon thereafter, to behave pretty wildly. Our first thought is that we have simply taken h too large, but reducing h only delays the 'explosion' and beyond a certain point (depending on the accuracy of our computer) any further reduction in h will not even do this. The magic words 'well behaved' which seemed so powerful to the apprentice sorcerer when things were going well, turn out to be useless in the face of real difficulties. What has gone wrong?

The answer is rather subtle and requires the result of the next exercise.

Exercise 9.3.4. **[A simple difference equation]** *Suppose that p and q are unequal non-zero numbers and write*

$$(t - p)(t - q) = t^2 + bt + c.$$

---

[7] 'It may be said that the fact makes a stronger impression [on the student if they perform the experiment themselves]. I say that this ought not to be the case. If the [student] does not believe the statements of his tutor – probably a clergyman of mature knowledge, recognised ability and blameless character – his suspicion is irrational and manifests a want of the power of appreciating evidence, a want fatal to his success in that branch of science which he is supposed to be cultivating.' Todhunter *The Conflict of Studies.*

We wish to study the sequence $u_0, u_1, u_2, \ldots$ given by

$$u_n + bu_{n-1} + cu_{n-2} = 0$$

(that is to say, by $u_n = -bu_{n-1} - cu_{n-2})$ for $n \geq 2$.

(i) Show that, if u_0 and u_1 are fixed, the remaining u_n are completely determined.

(ii) Show that $p^2 + bp + c = 0$ and deduce that $p^n + bp^{n-1} + cp^{n-2} = 0$ for all values of n.

(iii) Show that, if A and B are constants and $v_n = Ap^n + Bq^n$, then

$$v_n + bv_{n-1} + cv_{n-2} = 0$$

for all n.

(iv) Show that, given u_0 and u_1, we can find A and B such that

$$A + B = u_0$$
$$Ap + Bq = u_1.$$

(v) Deduce that the solutions of

$$u_n + bu_{n-1} + cu_{n-2} = 0$$

are given by $u_n = Ap^n + Bq^n$ for suitable constants A and B.

Armed with the result of the previous exercise, we look at the direct midpoint method applied to the system $y'(t) = -Ky(t)$, where $K > 0$, with $y(0) = 1$ and step length h. We wish to solve

$$y_{r+2} - y_r = 2hf(y_{r+1}, (r+1)h)$$

which, in this case, takes the form

$$y_{r+2} + 2Khy_{r+1} - y_r = 0.$$

By Exercise 9.3.4, we know that this yields

$$y_r = Ap^r + Bq^r,$$

where p and q are the roots of

$$t^2 + 2Kht - 1 = 0.$$

Exercise 9.3.5. *Show that, if we take $p > 0$,*

$$p = -Kh + \sqrt{1 + K^2h^2}, \quad q = -p^{-1}.$$

Thus if $y_0 = 0$, $y_1 = p$ and we solve our equations *exactly*, we get

$$y_r = p^r.$$

If $T > 0$, $h = T/N$ and N is large we get

$$y_N = \left(- Kh + \sqrt{1 + K^2 h^2}\right)^N \approx (1 - Kh)^N = \left(1 - \frac{KT}{N}\right)^N \approx e^{-KT}$$

just as we wanted.

Exercise 9.3.6. • *In fact more is true. Show that, there is a constant M such that*

$$\left| 1 + \frac{K^2 h^2}{2} - \sqrt{1 + K^2 h^2} \right| \le M h^4$$

provided h is small.

Show also that, if $|x|$, $|y| \le 1$, then $|x^N - y^N| \le N|x - y|$.

Now use Exercise 6.3.5 to show that there is a constant C such that, provided that h is small,

$$|y_N - e^{-KT}| \le C h^2,$$

so we do have the improvement in accuracy suggested.

However, in practice, we cannot expect to have $y_1 = p$ exactly and, in any case, we can only work to some limited accuracy, so what we expect to see is

$$y_r \approx A p^r + B(-p)^{-r},$$

where A is very close to 1 and B is very close to, *but not exactly equal to* 0. As r increases, the term $A p^r$ (which corresponds to the desired solution) will decrease towards 0 whilst the term $B(-p)^{-r}$ will increase rapidly in magnitude and eventually destroy any resemblance to our desired result.

Exercise 9.3.7. *Let p be as in our discussion. If $T > 0$, $h = T/N$ and N is large, show that*

$$(-p)^{-N} \approx (-1)^N e^{KT}.$$

The problem lies, not in the local error, but in the system of equations we have set up. The unwanted $B(-p)^{-r}$ is often referred to as a *parasitic* solution.[8]

The object of this section has been to sound a note of caution, not to induce a state of despair. The direct mid-point method is faster and more accurate

---

[8] It is worth observing that this shows that the 'estimation of the third term' undertaken on page 133 was a necessary exercise. The presence of a parasitic solution means that any attempt to produce a corresponding estimate for the direct mid-point method is doomed to failure.

than the Euler method in many cases[9] and the way ahead involves more use of Taylor's theorem to still further reduce the local error. However, no method can be applied blindly to all cases. We do not use the same methods to weigh an atom, a feather, an oil tanker and a star, and we cannot use the same method on all differential equations.

---

[9] Even so, I do not recommend using the direct mid-point method. There are several better alternatives.

10

Paradise lost

10.1 The snake enters the garden

In Section 6.3, we saw that, whatever the value of x, we can make

$$1 + \frac{x}{1!} + \frac{x^2}{2!} + \cdots + \frac{x^n}{n!}$$

as close to $\exp x$ as we like by increasing n. Why not just write

$$\exp x \underset{?}{=} 1 + \frac{x}{1!} + \frac{x^2}{2!} + \cdots + \frac{x^n}{n!} + \cdots$$

or, more generally,

$$f(x) \underset{?}{=} f(0) + \frac{f'(0)}{1!}x + \frac{f''(0)}{2!}x^2 + \cdots + \frac{f^{(n)}(0)}{n!}x^n + \cdots? \qquad \text{(A)}$$

We shall call this expression the *Taylor expansion* of f.

On the face of it, nothing could be simpler, but, to repeat Berkeley's words, '… if laying aside the Expressions we set ourselves attentively to consider the things themselves, which are supposed to be expressed or marked thereby, we shall discover much Emptiness, Darkness, and Confusion.'

Exercise 10.1.1. *Let* $F(x) = (1 + x)^{-1}$. *By computing* $F^{(n)}(x)$ *and then* $F^{(n)}(0)$, *obtain the Taylor expansion*

$$\frac{1}{1 + x} \underset{?}{=} 1 - x + x^2 - x^3 + x^4 + \cdots.$$

Setting $x = 2$ we obtain

$$\frac{1}{3} \underset{?}{=} 1 - 2 + 4 - 8 + 16 + \cdots.$$

The Duke of Wellington was greeted in the street by a man, saying, 'Mr Jones, I believe?' to which the Duke replied 'If you believe that, you will believe anything.'

Does this reflect problems with our earlier work? Suppose that we define the remainder term $R_n(f, x)$ by

$$R_n(f, x) = f(x) - \left(1 + \frac{f'(0)}{1!}x + \frac{f''(0)}{2!}x^2 + \cdots + \frac{f^{(n)}(0)}{n!}x^n\right)$$

so that

$$f(x) = f(0) + \frac{f'(0)}{1!}x + \frac{f''(0)}{2!}x^2 + \cdots + \frac{f^{(n)}(0)}{n!}x^n + R_n(f, x). \tag{B}$$

We showed in Section 6.2 that, if f is well behaved and $|f^{(n+1)}(t)| \leq M_{n+1}$ for $0 \leq t \leq x$, then

$$|R_n(f, x)| \leq \frac{M_{n+1}}{(n+1)!}|x|^{n+1}.$$

If M_{n+1} is very large, this may not give any useful information.

Exercise 10.1.2. *Let $F(t) = (1 + t)^{-1}$ and $x \geq 0$. Show that the smallest choice of M_{n+1} which gives $|F^{(n+1)}(t)| \leq M_{n+1}$ for $0 \leq t \leq x$ is $M_{n+1} = (n+1)!$. Our theorem thus gives*

$$|R_n(F, x)| \leq x^{n+1}.$$

The conclusion of Exercise 10.1.2 tells us that, in the example chosen, $R_n(F, x)$ does, indeed, become small if $0 \leq x < 1$ and n is large. What happens if $x \geq 1$? In this particular example we have a geometric series and

$$\frac{1}{1+x} = 1 - x + x^2 - x^3 + x^4 + \cdots + (-1)^n x^n + \frac{(-1)^{n+1}x^{n+1}}{1+x}.$$

Exercise 10.1.3. *Use simple algebra to verify the statement just made.*

Thus, if $F(x) = (1 + x)^{-1}$, we have

$$R_n(F, x) = \frac{(-1)^{n+1}x^{n+1}}{1+x}$$

and setting $x = 2$ in equation (B) gives us the unexciting news that

$$\frac{1}{3} = 1 - 2 + 4 - 8 + 16 + \cdots + (-1)^n 2^n + \frac{(-1)^{n+1}2^{n+1}}{3}.$$

It might be argued that the nonsense produced by considering the Taylor expansion of $F(x) = (1 + x)^{-1}$ at 2 has something to do with the fact that

$(1 + x)^{-1}$ behaves badly near -1, but this argument cannot be used for $G(x) = (1 + x^2)^{-1}$, which behaves well everywhere.

Exercise 10.1.4. *(i) By observing that $G(x) = F(x^2)$, show that, if $n = 2m$ or $n = 2m + 1$,*

$$G(x) = 1 - x^2 + x^4 - x^6 + x^8 + \cdots + (-1)^m x^{2m} + R_n(G, x),$$

where

$$R_n(G, x) = \frac{(-1)^{m+1} x^{2m+2}}{1 + x^2}.$$

Use Exercise 6.2.4 to show that $G^{(2r)}(0) = (-1)^r$ and $G^{(2r)}(0) = (-1)^r$ and write down the Taylor expansion of G.

(ii) Show that, if $|x| < 1$, we can approximate $G(x)$ arbitrarily well by taking sufficiently many terms in the Taylor expansion (that is to say, by taking n large enough), but that, if $|x| \geq 1$, this statement is false.

We now know that the speculative formula (A) produces nonsense for the well behaved function G when applied at points x with $|x| \geq 1$. There is nothing special about the value 1.

Exercise 10.1.5. *If $a > 0$, let us set*

$$G_a(x) = \frac{a^2}{a^2 + x^2}.$$

(i) Show that $G_a(x) = G(x/a)$.
(ii) Show that, if $n = 2m$ or $n = 2m + 1$,

$$G_a(x) = 1 - \frac{x^2}{a^2} + \frac{x^4}{a^4} - \frac{x^6}{a^6} + \frac{x^8}{a^8} + \cdots + (-1)^m \frac{x^{2m}}{a^{2m}} + R_n(G_a, x),$$

where

$$R_n(G_a, x) = \frac{(-1)^{m+1} x^{2m+2}}{a^{2m+2}} \frac{1}{a^2 + x^2}.$$

(iii) Show that, if $|x| < a$, we can approximate $G_a(x)$ arbitrarily well by taking sufficiently many terms the Taylor expansion (that is to say, by taking n large enough), but that, if $|x| \geq a$, this statement is false.

Remark. The fact that, whenever $a \neq 0$, we can find a well behaved function for which the speculative formula (A) produces nonsense at the point a raises the disturbing possibility that there exist well behaved functions for which the speculative formula (A) produces nonsense at all non-zero points. Such functions do indeed exist, but are outside the scope of this book.

Our next example of the problems associated with $\underset{?}{=}$ requires a result that is both interesting and uncontroversial. Recall that

$$\frac{1}{1+t} = 1 - t + t^2 - t^3 + \cdots + (-1)^{n-1}t^{n-1} + \frac{(-1)^n t^n}{1+t}.$$

Let $x \geq 0$. Integrating from 0 to x, we obtain

$$\log(1+x) = [\log(1+t)]_0^x = \int_0^x \frac{1}{1+t}\,dt$$

$$= \int_0^x \left(1 - t + t^2 - t^3 + \cdots + (-1)^{n-1}t^{n-1} + \frac{(-1)^n t^n}{1+t}\right) dt$$

$$= \int_0^x (1 - t + t^2 - t^3 + \cdots + (-1)^{n-1}t^{n-1})\,dt + T_n(x)$$

$$= \left[t - \frac{t^2}{2} + \frac{t^3}{3} - \frac{t^4}{4} + \cdots + (-1)^{n-1}\frac{t^n}{n}\right]_0^x + T_n(x)$$

$$= x - \frac{x^2}{2} + \frac{x^3}{3} - \frac{x^4}{4} + \cdots + (-1)^{n-1}\frac{x^n}{n} + T_n(x),$$

where

$$T_n(x) = (-1)^n \int_0^x \frac{t^n}{1+t}\,dt.$$

In order to produce something useful, we need to estimate $T_n(x)$. We observe that

$$0 \leq \frac{t^n}{1+t} \leq t^n$$

for $t \geq 0$ and so

$$0 \leq \int_0^x \frac{t^n}{1+t}\,dt \leq \int_0^x t^n\,dt = \left[\frac{t^{n+1}}{n+1}\right]_0^x = \frac{x^{n+1}}{n+1}.$$

Thus, if $0 \leq x \leq 1$, we have $|T_n(x)| \leq (n+1)^{-1}$ and

$$\log(1+x) \approx x - \frac{x^2}{2} + \frac{x^3}{3} - \frac{x^4}{4} + \cdots + (-1)^{n-1}\frac{x^n}{n}, \qquad \text{(C)}$$

where we can make the approximation as close as we please by taking n large enough. If we set $x = 1$, we get the particularly pleasing result

$$\log 2 \approx 1 - \frac{1}{2} + \frac{1}{3} - \frac{1}{4} + \cdots + (-1)^{n-1}\frac{1}{n}. \qquad \text{(D)}$$

Exercise 10.1.6. *Formula (D) is pleasing in an aesthetic rather than a computational sense. Show that we are far better off if we use formula (C) to approximate*

$\log 7/5$ *and* $\log 10/7$ *and then use the fact that* $\log 7/5 + \log 10/7 = \log 2$ *than if we use formula (D) directly. Can you suggest even better approaches?*

Exercise 10.1.7. *We do not need the result, but, for the sake of completeness, show that formula (C) fails if we choose* $x > 1$ *and* n *is large.*

Suppose that we now throw caution to the winds and write

$$\log 2 \underset{?}{=} 1 - \frac{1}{2} + \frac{1}{3} - \frac{1}{4} + \cdots + (-1)^{n-1}\frac{1}{n} + \cdots.$$

Then

$$\frac{1}{2}\log 2 \underset{?}{=} \frac{1}{2} - \frac{1}{4} + \frac{1}{6} - \frac{1}{8} + \cdots + (-1)^{n-1}\frac{1}{2n} + \cdots.$$

Placing the first expression above the second we have

$$\log 2 \underset{?}{=} 1 - \frac{1}{2} + \frac{1}{3} - \frac{1}{4} + \frac{1}{5} - \frac{1}{6} + \frac{1}{7} - \frac{1}{8} \cdots + (-1)^{4n-1}\frac{1}{4n} + \cdots,$$

$$\frac{1}{2}\log 2 \underset{?}{=} \quad \frac{1}{2} \quad - \frac{1}{4} \quad + \frac{1}{6} \quad - \frac{1}{8} \cdots + (-1)^{2n-1}\frac{1}{4n} + \cdots.$$

If we add the two expressions in the obvious manner, we obtain

$$\frac{3}{2}\log 2 \underset{?}{=} 1 + 0 + \frac{1}{3} - \frac{1}{2} + \frac{1}{5} + 0 + \frac{1}{7} - \frac{1}{4} + \cdots$$

which, after removing the zeros, gives us

$$\frac{3}{2}\log 2 \underset{?}{=} 1 + \frac{1}{3} - \frac{1}{2} + \frac{1}{5} + \frac{1}{7} - \frac{1}{4} + \cdots.$$

Exercise 10.1.8. *(i) Write out the formulae explicitly as far as* $1/16$.
(ii) Check that the $3k + 1$*st term on the right hand side of the expression just obtained (after removing zeros) is* $1/(4k + 1)$*, that the* $3k + 2$*th term is* $1/(4k + 3)$ *and the* $3k + 3$*th term is* $-1/(2(k + 1))$*. Conclude that the terms of the expression are the numbers* $(-1)^{r-1}/r$ *with each of these numbers occurring exactly once.*

Thus, if we add the numbers $(-1)^{r-1}/r$ in one order, we get $\log 2$, but, if we add them in another order, we get $\frac{3}{2}\log 2$. Earlier, we saw that using $\underset{?}{=}$ could lead to nonsense. Here it appears to lead to a contradiction. Why not just abandon the idea?

10.2 Too beautiful to lose

There were excellent reasons why the mathematicians of the early nineteenth century were unwilling to reject infinite sums. Their predecessors had used infinite sums in all sorts of situations and had obtained many striking results which certainly appeared to be true and which seemed inaccessible except by using this method.

As an example, let us consider the solution of differential equations. I shall explain the method in a case where we know the result by other means. Let us try to solve the differential equation

$$f''(t) + f(t) = 0.$$

We *assume* that we can expand f as an infinite sum

$$f(t) \underset{?}{=} a_0 + a_1t + a_2t^2 + a_3t^3 + \cdots .$$

We now *assume* that we can differentiate the infinite sum in the same way as we differentiate a polynomial so that

$$f''(t) \underset{?}{=} (1 \times 2)a_2 + (2 \times 3)a_3t + (3 \times 4)a_4t^2 + \cdots$$

and so the equation $f(t) = -f''(t)$ gives us

$$a_0 + a_1t + a_2t^2 + a_3t^3 + \cdots \underset{?}{=} -(1 \times 2)a_2 - (2 \times 3)a_3t - (3 \times 4)a_4t^2 - \cdots .$$

We now *assume*[1] that we can equate coefficients in the same way as we do for polynomials to obtain

$$a_0 = -(1 \times 2)a_2$$
$$a_1 = -(2 \times 3)a_3$$
$$a_2 = -(3 \times 4)a_4$$
$$a_3 = -(4 \times 5)a_5$$

$$\vdots$$

$$a_{n-2} = -\big((n-1) \times n\big)a_n$$

$$\vdots$$

---

[1] However, this assumption is relatively harmless. It bears the same relation to the other assumptions as petty theft does to grand larceny.

These equations yield

$$a_{2n} = \frac{-1}{(2n-1) \times 2n} a_{2n-2} = \frac{1}{(2n-3) \times (2n-2) \times (2n-1) \times 2n} a_{2n-4}$$

$$= \frac{-1}{(2n-5) \times (2n-4) \times (2n-3) \times (2n-2) \times (2n-1) \times 2n} a_{2n-6}$$

$$= \cdots = \frac{(-1)^n}{(2n)!} a_0.$$

Exercise 10.2.1. *Check, similarly, that*

$$a_{2n-1} = \frac{(-1)^{n-1}}{(2n-1)!} a_1.$$

If we write $A = a_0$, $B = a_1$, then the results just obtained give

$$f(t) \underset{?}{=} A + Bt - \frac{A}{2!}t^2 - \frac{B}{3!}t^3 + \frac{A}{4!}t^4 + \frac{B}{5!}t^5 - \frac{A}{6!}t^6 - \frac{B}{7!}t^7 + \cdots.$$

Assuming (in spite of the worrying example which closed the last section) that we can change the order of terms, this gives

$$f(t) \underset{?}{=} A\left(1 - \frac{t^2}{2!} + \frac{t^4}{4!} - \frac{t^6}{6!} + \cdots\right) + B\left(t - \frac{t^3}{3!} + \frac{t^5}{5!} - \frac{t^7}{7!} + \cdots\right).$$

We now recognise the Taylor expansions for cos and sin (if you do not recognise them, do Exercise 6.3.12 (i)) and obtain

$$f(t) \underset{?}{=} A\cos t + B\sin t.$$

Exercise 10.2.2. *Verify that, if $f(t) = A\cos t + B\sin t$, with A and B constants, then, indeed, $f''(t) + f(t) = 0$.*

A series of unjustified *assumptions* has produced a *correct* answer.

Exercise 10.2.3. *Solve the differential equation $g'(x) = -2xg(x)$ in the manner just given. Identify the Taylor series thus produced and hence find a simple formula for g. Check the formula by verifying that it satisfies the original equation.*

However, as Berkeley observed, 'when the Conclusion is evident and the Premises obscure, or the Conclusion accurate and the Premises inaccurate, we may safely pronounce that such Conclusion is neither evident nor accurate, in virtue of those obscure inaccurate Premises or Principles; but in virtue of some other Principles which perhaps the Demonstrator himself never knew or thought of.'

Exercise 10.2.4.• *This is a slightly artificial example but worth pondering. Show that, if we assume that our infinite sums can be manipulated like polynomials,*

$$(a_0 + a_1 x + a_2 x^2 + \cdots)(b_0 + b_1 x + b_2 x^2 + \cdots) \underset{?}{=} (c_0 + c_1 x + c_2 x^2 + \cdots)$$

with

$$c_r = a_0 b_r + a_1 b_{r-1} + a_2 a_{r-2} + \cdots + a_r b_0.$$

If f satisfies the differential equation $f'(x) = -2x f(x)^2$, and

$$f(x) \underset{?}{=} a_0 + a_1 x + a_2 x^2 + \cdots$$

write down the equations for the a_n and show that they are satisfied by

$$a_{2n-1} = 0, \quad a_{2n} = (-1)^n B^{n+1},$$

where B is a constant.
   We thus have

$$f(x) \underset{?}{=} B - B^2 x^2 + B^3 x^4 - \underset{?}{=} \frac{B}{1 + Bx^2}.$$

If $B > 0$ and we set $A = B^{-1}$, we obtain the result $f(x) \underset{?}{=} (A + x^2)^{-1}$. Check directly that this is a solution valid for all x. However, we have already agreed that the infinite sum in our expression makes no sense if $x \geq A^{1/2}$.

The differential equations we have just studied have solutions which are known functions, so we can verify the results directly. However, by the beginning of the nineteenth century, physics was beginning to produce a wide variety of differential equations of the form

$$f''(x) + a(x) f'(x) + b(x) f(x) = q(x),$$

where a, b and q are specified well behaved functions and we wish to know f. In many cases it is possible to find Taylor series for f, but these Taylor series are not associated with known functions. How can one decide if the Taylor series represent proper solutions rather than nonsense?

Taylor series were used over and over again by Laplace in his investigation of the stability of the solar system. Infinite sums of a different type were used by Fourier to investigate the laws of heat conduction. Cauchy needed Taylor series in his attempt to extend the calculus to the complex numbers. Dirichlet required yet another type of infinite sum (closely related to the sum

$$1 - \frac{1}{2} + \frac{1}{3} - \frac{1}{4} + \cdots + (-1)^{n-1} \frac{1}{n} + \cdots,$$

which worried us earlier by changing its value when we changed the order of its terms) for his profound study of the properties of prime numbers.

Looking back at the work of their predecessors, the reformers could see many results obtained by using infinite sums that could be shown to be true by other means. They could also find many cases where the use of infinite sums produced nonsense and a few cases, some rather artificial, but others not, where the result made sense but was actually false.

It is doubtful if any of them knew about Berkeley's criticism and, if they had known, it is doubtful if they would have cared. Although the 'old calculus' described in this book has gaps which make it vulnerable to the attacks of philosophers and angry bishops, it is a tool that will not fail in the hands of a competent practitioner. However, the mass of 'truths', 'probable truths', 'possible truths' and 'clear falsehoods' accumulated in the study of infinite sums threatened to block any further progress.

We could compare the old calculus to a battered old car which will potter along quite safely provided you do nothing stupid and the new calculus, as it was in 1800, to a sports car with the unfortunate fault that both its brakes and steering are liable to fail without warning. The only way out was to recast the theory of infinite sums on a rigorous basis with exact definitions, precise statements of theorems and cast-iron proofs.

The leader in this movement was Cauchy.[2] The brilliant young Norwegian mathematician, Abel, wrote back from Paris to his old teacher, Holmboe:- 'Cauchy is mad and it is impossible to get along with him, although he is the one person who knows how to do mathematics today. He writes excellent things, but very mixed up. I understood almost nothing of what he meant at first but I am beginning to comprehend him.'[3] He then writes of his own work on convergence and how '... the scales fell from my eyes ... [and I saw that] apart from the simplest cases, such as the geometric series, there is hardly an example of an infinite series whose sum has been determined rigorously, that is to say that the most important part of mathematics has no foundation. It is true that most of the results obtained are correct, but this fact is itself most peculiar.'

However, within less than 40 years, thanks in part to Abel's own work, French and German mathematicians could manipulate infinite sums correctly

---

[2] A truly great mathematician who worked on a wide range of subjects and 'touched nothing he did not adorn', but a difficult man.

[3] Letter numbered XX in the second edition of Abel's *Collected Works*. In the first edition, the first sentence simply ran 'Cauchy is the one person who knows how to do mathematics today.' The phrase 'very mixed up' ('trés brouillé') was replaced by 'lacks clarity' ('manque de clarté').

and freely.[4] The reader may be interested to learn that, in the new analysis, both of the expressions

$$\log 2 = 1 - \frac{1}{2} + \frac{1}{3} - \frac{1}{4} + \frac{1}{5} - \frac{1}{6} + \cdots$$

$$\frac{3}{2}\log 2 = 1 + \frac{1}{3} - \frac{1}{2} + \frac{1}{5} + \frac{1}{7} - \frac{1}{4} + \cdots$$

are correct and that altering the order in which we sum an infinite series can, indeed, change the final sum.

---

[4] The same could not be said of British mathematicians whom Hardy describes as exhibiting 'a singular and often entertaining mixture of occasional shrewdness and fundamental incompetence.' (From Hardy's *Divergent Series*.)

11

Paradise regained

11.1 A short pep talk

The calculus, as presented in this book, is a collection of methods whose justification depends on appeals to intuition. As we saw in the previous chapter, this causes serious problems when we attempt to push these methods past a certain point. It is part of the mathematician's ethos to push methods as far as they will go, so it is important to reconstruct calculus without any appeals to intuition.

Fortunately, the nineteenth-century reformers were able to perform the reconstruction in such a way as to retain the main methods and results of the old calculus. Sometimes, people claim that, since the reconstruction was so successful, it must have been unnecessary. However, mathematicians like Gauss, Cauchy, Dirichlet and Weierstrass did not reconstruct analysis with a view to the past, but with a view to the future. The questions they wished to ask and the methods they wished to use are simply inexpressible in the language of the old calculus.[1]

Before starting the reconstruction, the reader should place this book on the nursery shelf next to *The Very Hungry Caterpillar* and *My First Counting Book*, since any attempt to mix the 'old calculus' with the 'new analysis' is likely to be disastrous for the reasons explained by Abel in another letter to Halsteen.

> Very few theorems in advanced calculus have been proved completely rigorously. Everywhere we find the unfortunate habit of assuming that what is true in a few examples is true in all cases. It is truly remarkable how we rarely see what are called paradoxes. It would be very interesting to find out why this is the case. In my opinion, it is because most of the functions studied so far are given by their Taylor series. When we need other types of functions, although it must be admitted this

---

[1] Many beautiful results were obtained within the framework of the old calculus, but almost all the important advances since 1900 require the new analysis.

only happens occasionally, everything falls to pieces and it is easy to derive false results giving rise to an endless chain of lying propositions. I have examined several of these and found pleasure in exposing the underlying fallacies. When one uses the proper general method this is not difficult; but I have had to proceed with great caution because once I have accepted a proposition without a rigorous proof (that is to say, without any proof whatsoever) it takes such root in my mind that, at any moment, there is a danger that I will use it without examining it later.

Letter XXI in Abel's *Collected Works*. (My translation is very free.)

I shall not attempt to provide a first course in the new analysis (and, in particular, I shall not prove any theorems), but I shall discuss some of the underlying issues. Please bear in mind that the discussion has a British accent (see page 3).

11.2 The Euclidean method

Mathematics is very difficult and mathematicians frequently make mistakes. Sometimes these are trivial misprints and misstatements. Sometimes they are more substantial errors which can, none the less, be corrected. Sometimes the mistakes are so serious that nothing can be saved from the resulting mess. Why should you suppose that what your lecturer says is not mistaken?

The first thing that you will notice about most advanced mathematical presentations is that they consist of a series of numbered Definitions, Propositions, Lemmas, Theorems and so on.[2] This is because the lecturer is using the Euclidean method.

Ever since a group of apes descended from the trees, they have been arguing amongst themselves. The first method of argument, which is still very common, involves shouting very loudly and, in extreme cases, throwing large stones. The ape with the loudest voice or the biggest stone is deemed to have won.

Later, people invented the method of advocacy. One side puts forward all the arguments it can think of ('I was not there. If I was, I did not do it and, if I did it, I am very sorry.') and then the other side puts forward all of its arguments and then a judge or the electorate decides who had the better arguments. Odd though this system is, it is often the best we have.

In certain circumstances we can use experiment. If two physicists disagree, they can agree on an experiment which distinguishes between their two theories. (If no such experiment exists, they are not discussing physics, but something else.) If experiment decides against a theory, it does not matter how rigorous the mathematics behind it is, the theory must be abandoned. If the theory survives

---

[2] It is said that Professor X's lecture notes included Joke 1.1.9.

the test then, again, the rigour or otherwise of the underlying mathematics is essentially irrelevant. There are excellent reasons why engineers and physicists are untroubled by foundational mathematics. On the other hand, we cannot argue that the calculus is correct because bridges built using it do not fall down, since this would mean that every time a bridge fell down it would cast doubt on the calculus.

The final type of argument was perfected by Euclid and has been used by mathematicians ever since. It starts with clear definitions and a clear statement of assumptions (that is to say, *axioms*) and then sets out a series of clear assertions. Each assertion (that is to say, *theorem*) must be justified by an argument which uses only the initial assumptions and any previously proved assertion. If you fail to prove a particular assertion, then it and any assertions whose purported proof depends on it, remain unproved.

Important note. It is very useful to observe that Euclidean arguments can be read backwards as well as forwards. We start with the final conclusion Z and observe that it will follow if we can prove Y and Y will follow if we can prove X and so on backwards until we reach assumption A. It is often easier to grasp a Euclidean argument in this way.

It is not hard to find eminent people who object to the Euclidean method as unsuitable for the purposes of teaching and research. These are not the purpose of the Euclidean method which is to *test correctness* and which is a very powerful tool for this purpose.[3] The standard first course in analysis, like most advanced courses in mathematics, is set out in Euclidean form so that those attending the course can see the theory being put to a proper test.

A lot of people (including many of the participants) look at a first analysis course and see a lecturer writing notes on a blackboard that the students copy down. The students then take the notes away and store them carefully until the week before the exam. The notes are then revised by highlighting them until at least 25% of their surface area will glow in the dark. There is then the exam itself, after which the papers are marked according to how many of the lecturer's original words can be glimpsed through the garbled murk. Once the grades have been announced, both lecturer and students can go back to their real lives.

Others see the first analysis course as a joint enterprise of students and teacher in which they examine the justification for the calculus, link by link, looking for weaknesses. If they succeed in finding a serious flaw, then the

---

[3] I also believe, contrary to the eminent people just referred to, that the Euclidean method is a very useful tool both for teaching and research, though not the only one.

participant who has found that flaw will enjoy instant fame throughout the mathematical community. If they find a minor flaw, then it is probably an error of the lecturer and both the lecturer and students will benefit from seeing how it is corrected.

There is an element of truth in both of the views I have recorded. However, you will enjoy and benefit from the course much more if you act as though the second view is correct.

Bertrand Russell thought very little of the mathematics teaching he received in Cambridge,[4] but recorded that:

> I cannot remember any instance of a teacher resenting it when one of his pupils showed him to be in error, though I can remember quite a number of occasions on which pupils succeeded in performing this feat. Once during a lecture on hydrostatics, one of the young men interrupted to say: 'Have you not forgotten the centrifugal forces on the lid?' the lecturer gasped and then said: 'I have been doing this example this way for twenty years, but you are right.'
>
> Russell *The Autobiography of Bertrand Russell*

11.3 Are there enough numbers?

Naturally, the course begins with a set of definitions of which the most important is that of a limit.[5] Sometimes the lecturer attempts to show that these definitions are natural, but, although such explanations may be interesting, they do not form part of the theory. The reason why mathematicians use these definitions is that it turns out that they give a useful and coherent theory. We know that, before Cauchy, other mathematicians tried other routes towards a rigorous calculus which were less successful and that, during the twentieth century, mathematicians produced other different, but successful, theories of analysis.[6]

In many ways, the definitions in analysis parallel the rules governing the movements of pieces in chess. A bishop can move any number of squares along any diagonal on which it sits, not because it is 'natural' for a bishop to move in this way, but because this is the definition of a bishop. Once the lecturer and her class have agreed on the definition of a continuous function, any further

---

[4] 'It was only after I left Cambridge and began to live abroad that I discovered what I should have been taught.'

[5] Although the $o(h)$ arguments of this book can be restated in terms of limits, the reader is strongly advised to abandon this seductive notation for the duration of her studies.

[6] Since they must all contain versions of the 'old calculus', they all bear a strong family resemblance, but 'constructive analysis' and 'non-standard analysis' are genuinely different – cousins of the standard theory rather than siblings.

properties of a continuous function must be deduced by clear steps from the definition chosen.

Important note. Because definitions 'define what they define' and not what we think they define, it is important to have a store of examples of things which obey a definition and things which do not obey the definition. If you do not possess such a store you cannot be said to have understood the definition.

Once the definitions have been laid down, the lecturer will prove a set of more or less easy results, but simply proving a lot of results does not bring us much closer to justifying the calculus. If we go no further, we are in danger of imitating Victor Borge's uncle who invented the cure for which there was no disease.[7] The new analysis is *not*, as is sometimes said, 'epsilon-delta calculus' (in other words, the study of limits), but should properly be thought of as the study of the real numbers.[8]

I have said repeatedly that the calculus has gaps, but I have rather glossed over[9] what they are. In my opinion, they have to do with *existence* and *uniqueness*, that is to say, the assumption that the solution to a given problem always *exists* and that it is *unique*.

Throughout this book, we have assumed the existence of numbers like $\sqrt{2}$ defined to be the positive root of the equation

$$x^2 = 2$$

and $e = \exp 1$ defined to be the solution of

$$\log e = 1.$$

In Shakespeare's *Henry IV, Part I*, Glendower boasts that 'I can call spirits from the vasty deep' and Hotspur replies 'Why, so can I, or so can any man; But will they come when you do call for them?' What right have we to make such definitions?

One possible answer is that this 'postulational method' has been very successful in the past. Negative numbers, surds (that is to say, numbers like $\sqrt{2}$) and imaginary numbers have all entered mathematics by this route. People simply assumed (sometimes with confidence, more usually without[10]) that they could exist and investigated their properties, discovering in the process that they were useful and did not seem to cause paradoxes.

---

[7] Unfortunately, he died of it.

[8] If the reader wishes, she may replace the term 'real numbers' by 'numbers' or 'the numbers employed in the theory of the calculus' whenever I use it.

[9] As in: 'He tried to sell me the Brooklyn Bridge, but rather glossed over the question of ownership.'

[10] Consider the non-technical meanings of negative, irrational, imaginary and absurd.

However, this 'might makes right' justification is not entirely convincing. It is true that there is a theory in which the equation

$$x^2 + 1 = 0$$

has a solution, but there is no theory in which this is the case and the standard rules for inequalities hold.[11] When we postulate the existence of a certain number, how can we sure that it obeys the same rules as other numbers?

Things can get worse than this. If we assume that we can solve $\exp x = 1$, why should we not assume that we can solve $\exp y \underset{?}{=} 0$? Regrettably, we then get

$$1 = \exp 0 = \exp(y - y) = \exp y \exp(-y) \underset{?}{=} 0 \exp(-y) = 0,$$

which is not acceptable.

It turns out that many of the numbers which we require will exist automatically if the following theorem is valid.

Theorem 11.3.1. **[The intermediate value theorem]** *If f is continuous, $a < b$ and $f(a) \le 0 \le f(b)$, then there exists a number c with $a \le c \le b$ such that $f(c) = 0$.*

Exercise 11.3.2. *(i) Apply the intermediate value theorem to $f(t) = t^2 - 2$, with $a = 0$, $b = 2$.*

(ii) Recall from Exercise 3.1.5 that there exists an integer N such that $\log N \ge 2$. Apply the intermediate value theorem to $f(t) = \log t - 1$, with $a = 1$, $b = N$.

Exercise 11.3.3. *Show that there always exist a pair of diametrically opposite points on the earth's equator with the same temperature.*

An early nineteenth-century mathematician might be tempted to say that the notion of a continuous function f is such that, as t increases from a to b, $f(t)$ must necessarily pass through all values intermediate between $f(a)$ and $f(b)$. After all, the *Oxford Dictionary* says that continuous means 'Characterised by continuity; extending in space without interruption of substance; having no interstices or breaks; having its parts in immediate connection; connected, unbroken.' Unfortunately we have chosen the phrase 'continuous function' to

---

[11] Observe that the standard rules tell us that the square of a number must be a positive number.

mean 'changing very little at a sufficiently fine scale' and it is not at all clear why this should imply 'passing through all values'.

If we try to prove the theorem by 'lion hunting' (see page 109), we discover that our proof requires the real numbers to have a rather natural property.[12] We isolate this property, give it the name *The Fundamental Axiom of Analysis* and use it to prove the intermediate value theorem.

One way of expressing the Fundamental Axiom is to say that 'An increasing sequence bounded above tends to a limit.' From our point of view, this amounts to saying that there are *sufficiently many* real numbers to provide every bounded increasing sequence with a limit.[13] The intermediate value theorem can be thought of as saying that there are *sufficiently many* real numbers for every continuous function taking both positive and negative values to have a zero.

Exercise 11.3.4. *Go back through the book looking for points where we have used the intermediate value theorem.*
[Hint: There are examples on pages 48, 54, 59, 109 and on many other pages.]

Before the reader objects that the Fundamental Axiom of Analysis is itself a postulate, she should recall the well-known story (which goes back, in various versions, at least 200 years) of a young academic arguing with someone who claims that the earth rests on the back of an elephant. 'But what does the elephant stand on?' 'It stands on a turtle.' 'And what does the turtle stand on?' 'It stands on another turtle.' 'And what does that turtle stand on?' 'Very clever, young man, very clever, . . ., but it is turtles all the way down.'

Every mathematical theory must start from a certain number of assumptions. The first course in analysis seeks to show that the *entire* theory of the calculus follows from the *single* Fundamental Axiom (plus the usual rules of arithmetic and the usual rules of inference). The calculus and all its developments (complex analysis, harmonic analysis, differential geometry, advanced probability, . . .) rest like a giant inverted pyramid on the Fundamental Axiom.

No one *has* to accept the Fundamental Axiom,[14] but, if you accept the Axiom, then, it is claimed, you must accept the calculus and all its consequences.

---

[12] The statement that the property is natural is rhetorical rather than mathematical.
[13] It also implies (though this is by no means obvious) there are *sufficiently few* real numbers to allow every bounded increasing sequence of real numbers to have a limit. The need for an equivalent result was recognised by the Ancient Greeks and it is now called the 'Axiom of Archimedes'. We make use of this 'Axiom' in Exercise 2.2.11 and elsewhere when we claim, in effect, that if $|a - b| < 1/n$ for every strictly positive integer n, then $a = b$.
[14] Except temporarily, to pass first year exams.

11.4 Can we guarantee a maximum?

The next important theorem in the standard treatment of analysis deals with
the existence of maxima. It is clear that, when discussing hills and dales and
elsewhere in the book, we have assumed that certain functions have maxima
without giving any proof.

The reader may be inclined to ask if this really matters. After all, we could
always replace the word 'maximum' by the phrase 'maximum, if it exists'.
However, in the hands of Dirichlet and Riemann, the study of maxima proved
to be a powerful tool for the discovery and apparent proof of remarkable new
theorems. Unfortunately, Weierstrass showed that in certain circumstances the
maxima required did not exist and so the proofs were incomplete.

This unpleasant surprise made mathematicians extremely cautious in dealing
with maxima. The next few observations show the kind of precautions that were
necessary.

Observation 1. If we take $f(x) = x$, then there is no K such that $f(x) \leq K$
for all x. We see that the continuous function f is unbounded.

Observation 2. If we take

$$f(x) = \frac{x^2}{1 + x^2}$$

then $0 \leq f(x) \leq 1$, but, given any x, we can find a y with $f(x) < f(y)$. We
see that the continuous function f is bounded but has no global maximum.

Observation 3. If we restrict our attention to those x with $0 < x < 1$ and set
$f(x) = 1/x$, then there is no K such that $f(x) \leq K$ whenever $0 < x < 1$. We
see that the continuous function f is unbounded on the interval from 0 to 1
excluding the end points.

Observation 4. If we restrict our attention to those x with $0 < x < 1$ and set
$f(x) = x$, then $0 \leq f(x) \leq 1$ whenever $0 < x < 1$, but given any $0 < x < 1$,
we can find a y with $0 < y < 1$ and $f(y) > f(x)$. We see that the continuous
function f is bounded on the interval from 0 to 1, excluding the end points, but
does not have a global maximum.

Exercise 11.4.1. *Check the statements just made. Show also that the functions
considered have no local maxima.*

However, the following theorem is true.

Theorem 11.4.2. *Let $a < b$ and let us restrict our attention to those points x
with $a \leq x \leq b$. If f is a continuous function over the range considered, then*

there exist points y_1 and y_2 such that $a \le y_1 \le b$, $a \le y_2 \le b$ and

$$f(y_1) \le f(x) \le f(y_2)$$

whenever x with $a \le x \le b$.

Thus a continuous function on a bounded interval including its end points has a global maximum and a global minimum.

It is usual to deduce Theorem 11.4.2 in two steps, first showing that f is bounded and then showing that it has global maxima and minima. Both results can be deduced in a natural manner from the Fundamental Axiom. Theorem 11.4.2 can be interpreted as saying that there are enough real numbers to provide the appropriate maximum.

11.5 A glass wall problem

Students often think that problems which take a great many words to state must be harder than problems which can be stated briefly. However, the longer question often contains more clues as to its solution.

In the same way, technical difficulties associated with a research problem may actually point the way to a solution. Often the really hard problems in mathematics are very simply stated and are hard because their statement is so simple. I think of such problems as *glass wall* problems because they present no handhold to a possible climber.

In the previous two sections, we discussed existence problems that arose in trying to rigorise calculus. However, there is also a very serious uniqueness problem.

When we discussed the fundamental theorem of the calculus, we said that, if f is continuous, then the solutions of the differential equation

$$g'(t) = f(t)$$

are given by

$$g(t) = \int_a^t f(x)\,dx + c, \qquad \bigstar$$

where c is a constant. However, although it is certainly true that, if g is defined by equation $\bigstar$, then it satisfies the differential equation $g'(t) = f(t)$, it is by no means obvious[15] that functions of this form are the *only* solutions.

---

[15] In spite of the hand-waving assurances on page 43.

Let us write down what we wish to be true.[16]

Plausible Statement A. If f is continuous, then $g'(t) = f(t)$ implies $g(t) = \int_a^t f(x)\,dx + c$ for some constant c.

If we consider the particular case when $f(t) = 0$, we obtain an even simpler plausible statement.

Plausible Statement B. If $g'(t) = 0$ for all values of t, then $g(t) = c$ for some constant c (i.e. g is a constant function).

Exercise 11.5.1. *Suppose that Plausible Statement B is true. If f is continuous and $g'(t) = f(t)$ set*

$$G(t) = g(t) - \int_a^t f(x)\,dx.$$

By looking at $G'(t)$, show that

$$g(t) = \int_a^t f(x)\,dx + c$$

for some constant. Thus Plausible Statement B implies Plausible Statement A.

As it stands, the proof of Plausible Statement B appears to represent a perfect glass wall problem. On the one hand, it seems obvious that a function of zero slope must be constant. On the other hand, every attempt to prove it by saying that, if something is approximately linear with zero slope, where the approximation improves as the scale gets finer, then it must be constant, seems to fail.

If we think long enough, it may become clear that Plausible Statement B is linked to another plausible statement which the reader will recognise as an old friend.

Plausible Statement C. If $|g'(t)| \le M$ for all t, then $|g(b) - g(a)| \le M|b - a|$ for all a and b.

Exercise 11.5.2. *Obtain Plausible Statement B from Plausible Statement C by setting $M = 0$.*

At this point the reader may scratch her head. Have we not already proved Plausible Statement C in Section 2.4? If she looks back, she will see that we did indeed prove Plausible Statement C, but we did so by assuming Plausible Statement A!

---

[16] The reader will recognise elements of Sections 2.3 and 2.4 but, as I warned her there, these elements will be combined in a very different way.

Fortunately, it is possible to prove Plausible Statement C directly, though, as the reader may by now expect, the proof depends on exploiting the Fundamental Axiom.

Theorem 11.5.3. **[The mean value inequality]** *If* $|g'(t)| \leq M$ *for all* t, *then* $|g(b) - g(a)| \leq M|b - a|$ *for all* a *and* b.

For reasons of tradition as much as anything else, the mean value inequality is usually deduced from a slightly more powerful result known as the mean value theorem.

I strongly advise the reader to make a note of every time the mean value inequality (or some similar result) is used, since otherwise it is easy to lose sight of its importance. I also advise the reader to exercise great care whenever she proves the theorem, since it is fatally easy to produce circular arguments.

11.6 What next?

It is generally agreed that the intermediate value theorem, the theorem on the existence of maxima and the mean value theorem represent the centre of any first course in analysis. Such a course will also discuss and justify rules for dealing with infinite sums. It will also give a definition of an integral which does not depend on any 'intuition' concerning the properties of 'area'. It turns out that the Fundamental Axiom is also required to show that $\int_a^b f(t)\,dt$ exists whenever f is continuous,[17] but that the rest of elementary 'integral calculus' is easily justified. The fundamental theorem of the calculus is now proved in two stages. The proof that 'differentiation reverses integration' is, essentially, that given in this book and the proof that 'integration reverses differentiation' uses the mean value inequality as outlined in the previous section.[18]

There are many books that set out the contents of a first course in analysis. Burkill's *A First Course in Mathematical Analysis* [1] is elegant and to the point. If you have a good background in calculus, this may well be the book for you. If you are less well prepared or you prefer a more discursive style, then Spivak's *Calculus* [6] is excellent. Both books have good exercises.[19]

---

[17] We might expect this, since the intermediate value theorem shows that $\int_1^x (1/t)\,dt$ with $x > 0$ takes every real value as x varies.

[18] Some careful writers underline the point by referring to the results as the *first* and *second* fundamental theorems of the calculus, but this useful distinction has not caught on.

[19] These books have my strong recommendation, but it is best to discuss further reading with someone who knows you personally and choose a book which fits in with whatever your present or intended future educational institution does. If you want to widen your mathematical background, Courant and Robbins' *What is Mathematics?* [2] remains unsurpassed.

Once the first course in analysis is out of the way, we are free to advance towards the broad sunlit uplands of modern analysis and the rest of advanced mathematics.

11.7 The second turtle

Rigorous calculus based on the Fundamental Axiom was completely successful in banishing the paradoxes which threatened to block further progress. However, mathematicians being what they are, they set about looking for a second turtle on which to stand their first turtle. Recall that I said that rigorous calculus depended on the Fundamental Axiom (plus the usual rules of arithmetic and the usual rules of inference). If you look at the this statement sufficiently suspiciously, you may begin to wonder what 'the usual rules of inference' actually are.

Of course, this is a question about mathematics in general and not just analysis. The second turtle has to support not only the turtle of analysis, but the turtle of geometry, the turtle of arithmetic and any other turtles that mathematicians wish to study. It turns out that, in trying to make clear the rules of mathematical inference, we also have to make clear the nature of a mathematical object and draw up rules about how to construct one mathematical object from another. The study of mathematical inference is called *mathematical logic* and the study of how mathematical objects are constructed is called *set theory*. Analysts tend to find that set theory impinges rather more on their working practices than mathematical logic. Standard set theory itself rests on a number of axioms which are nicely set out in the classic text *Naive Set Theory* [3] by Halmos.

Starting from these axioms we can *construct* the positive integers, the rational numbers and the real numbers in such a way that the usual rules of arithmetic and the Fundamental Axiom hold. By constructing the real numbers rather than postulating their existence we seem to evade Russell's reproach[20] that 'The method of "postulating" what we want has many advantages; they are the same as the advantages of theft over honest toil.' However, in order to perform the constructions, we have to accept the axioms of standard set theory. The majority of mathematicians accept these axioms, not because they are obvious, but because no one can see any way to deduce them from something which is more obvious. A minority (though a respectable minority) seek a different second turtle but, at least for the moment, it seems unprofitable to seek a third turtle.

---

[20] In his *Introduction to Mathematical Philosophy*.

Further reading

[1] J. C. Burkill. *A First Course in Mathematical Analysis*. Cambridge University Press, Cambridge, 1962.

[2] R. Courant and H. Robbins. *What is Mathematics?* Oxford University Press, Oxford, 1941. OUP has issued this book in various forms including a second edition with an extra chapter by Ian Stewart discussing later developments.

[3] P. R. Halmos. *Naive Set Theory*. Van Nostrand, Princeton, 1960. The later publishing history is complicated, but the book is available in many libraries and is usually in print.

[4] F. Klein. *Elementary Mathematics from an Advanced Standpoint (Part 1)*. Macmillan, New York, 1932. Third edition. Translated by E. R. Hedrick and C. A. Hedrick. There is a Dover reprint.

[5] A. E. Maxwell. *An Analytical Calculus*, volume 1. Cambridge University Press, Cambridge, 1957. There are three further volumes covering successively more advanced topics.

[6] M. Spivak. *Calculus*. Cambridge University Press, Cambridge, third edition, 1994.

[7] S. P. Thompson. *Calculus Made Easy*. Macmillan, London, second edition, 1914. There are several editions revised by various hands. The revisions are not always happy. The 1914 edition is available from Project Gutenberg.

Index

Printed in the United States
By Bookmasters